ABOUT THE AUTHOR

A Physicist, Travis S. Taylor, PhD, has worked on various programs for the Department of Defense and NASA for the past sixteen years. His current work involves advanced propulsion concepts, substantial space telescopes, space-based beamed energy systems, and next generation space launch concepts.

His expertise in the field of quantum physics gives him the expertise to describe the real science behind The Law of Attraction: Travis holds a doctorate in optical science and engineering, three masters degrees in physics, aerospace engineering, and astronomy, and a bachelors degree in electrical engineering.

He lives in Harvest, AL with his wife Karen and their family.

Novels by Travis S. Taylor

One Day on Mars
The Tau Ceti Agenda
One Good Soldier

Warp Speed
The Quantum Connection

with John Ringo:
Von Neumann's War
Vorpal Blade
Manxome Foe
Claws That Catch

The Science Behind the Secret

Decoding
The Law of Attraction
&
The Universal Quantum
Connection

Travis S. Taylor, Ph. D.

Introduction by
John Edward

Afterword by
Stuart Hameroff

THE SCIENCE BEHIND THE SECRET

Copyright © 2010 by Travis S. Taylor

Permission is granted to copy, distribute and/or modify the following four images under the terms of the GNU Free Documentation License, Version 1.3 or any later version published by the Free Software Foundation; with no Invariant Sections, no Front-Cover Texts, and no Back-Cover Texts. A copy of the license is available at http://www.gnu.org/copyleft/fdl.html:

Map of Ancient Indus Valley Civilization in Figure 3.1: Copyright © 2005 by MM. (Source: http://en.wikipedia.org/wiki/File:Civilt%C3%A0ValleIndoMappa.png)

Photo of Gautama Buddha painting in Figure 3.5: Copyright © 2006 by Kirit Sælensminde. (Source: http://en.wikipedia.org/wiki/File:Sermon in the Deer Park depicted at Wat Chedi Liem-KayEss-1.jpeg)

Photo of stained-glass window of William of Ockham in Figure 3.13: Copyright © 2007 by Moscarlop. (Source: http://en.wikipedia.org/wiki/File:William of Ockham.png)

Photo of Stuart Hameroff in Figure 3.20: Copyright © 2008 by Nima Kasraie. (Source: http://en.wikipedia.org/wiki/File:Stuart Hameroff TASC2008.JPG)

Permission is granted to copy, distribute and/or modify the following two images under the terms of the Creative Commons Attribution ShareAlike 3.0 License. A copy of the license is available at http://creativecommons.org/licenses/by-sa/3.0/:

Photo of Tirthankar Mahavir Swami in Figure 3.2: Copyright © 2001 by Dayodaya. (Source: http://en.wikipedia.org/wiki/File:Mahavir.jpg)

Photo of Louise Hay in Figure 3.17: Copyright © 2008 by Heiko Antoni. (Source: http://en.wikipedia.org/wiki/File:LouiseHay.JPG)

A Baen Books Original

Baen Publishing Enterprises
P.O. Box 1188
Wake Forest, NC 27588
www.baen.com

ISBN 13: 978-1-4391-3339-2

Library of Congress Cataloging-in-Publication Data

Taylor, Travis S.
  The science behind The secret / by Travis S. Taylor.
    p. cm.
  "A Baen Books original."
  ISBN 978-1-4391-3339-2 (trade pbk.)
 1. Science--Miscellanea. 2. Quantum theory. 3. New thought. 4. Byrne, Rhonda. Secret. I. Title.
  Q173.T276 2010
  153.3--dc22
                                            2009050673

First printing, March 2010

Distributed by Simon & Schuster
1230 Avenue of the Americas
New York, NY 10020

Printed in the United States of America

10 9 8 7 6 5 4 3 2 1

Contents

Introduction by John Edward..................1

Chapter 1: Life, Hard Work, and Eureka Moments 5

Chapter 2: The Secret and the Law of Attraction 21

Chapter 3: Where Did The Secret and the Law of Attraction Come From? 31

Chapter 4: Why Einstein Hated Schrödinger's Cat 67

Chapter 5: There are Computers and then there are Quantum Computers 91

Chapter 6: "Brain and Brain! What is brain?" 99

Chapter 7: The Qwiff of Your Desire, or Maintaining Your Train of Thought 111

Chapter 8: The Universal Quantum Connection........................121

Chapter 9: Use Your Quantum Brain Wisely 137

Chapter 10: There Will Always be Skeptics So Don't Let them Get to You! 147

Chapter 11: The Proof is in the Pudding! 157

Chapter 12: Final Thoughts 173

Afterword by Stuart Hameroff 179

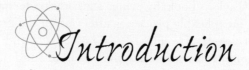

Introduction

For thousands of years there have been people who seem to have a "connection" with invisible forces, whether it's religious leaders, or medicine men, or even scientists; they seem to hold a special key to inside information.

This mysterious thread can be explained through Einstein's theory that everything is energy—but in various forms (including mass). We can talk about waves and particles on the scientific front, but it has an equal impact on the spiritual level: it's all part of a much larger information superhighway. The good news is that everyone has access to these wavelengths. So it's exciting when the areas of both the physical world and the spiritual world meet, and it is a significant moment when we can get a glimpse of the "wholeness" of the world, universe, and our lives.

In the work I do, it is all about "the energy." It's the information that I can tune into and decode or decipher and share with others. While it may seem remotely distant from science, it is still about energy.

Quantum physics is probably the best explanation of how "paranormal" things can happen: it's the *energy* that comprises

1

the invisible forces. Even though the two areas of thought—science and spirituality—may seem incongruous and in opposition, clearly they are not. It's just that most people take on religious beliefs as faith, and scientists seek empirical data. But both are really living with the same conclusion: everything is vibrating around us—it's a matter of tuning in.

In Travis Taylor's book, he brings a greater understanding of how these inexplicable processes affect us on the grandest and smallest levels. He illustrates how each and every important thought can influence the world around us. It is a validation and an encouragement to have an awareness to align ourselves with the highest possible vibration.

—**John Edward,**
September 2009

Chapter 1:

Life, Hard Work, and Eureka Moments

> "You create your own universe as you go along."
> —Winston Churchill

Every thought you have has an impact on the universe. That is what this book is about. Is it true? When you think of things do those thoughts have any real connection with the rest of the universe? I truly believe so, and I'm going to spend the pages of this book explaining to you why I believe this.

There are new discoveries about the brain, new theories about quantum physics, an understanding about how the universe came about (at least the known local universe), and a modern understanding of quantum computing that has led me to a conclusion that I'd have never thought I'd come to. *Your thoughts really do interact with the universe at the level of quantum physics.*

> "All that we are is the result of what we have thought. The mind is everything. What we think we become."
> —Buddha

I've never been a big believer in the various self-help philosophies out there proliferating across the globe. And, to tell the truth, I used to be quite skeptical about all those "power of positive thinking" people. But heck, I grew up in north Alabama, the Deep South of the United States, and it is a well-known phenomenon to all Southern grandmas that if something bad hasn't happened lately, then it is only a matter of time before it does. Sometimes it seems that my grandma's and my parents' generation get the happiest when a disaster occurs, so they can spring into action to talk about it and gossip and to cook everything in the house! If it had been a while since the last friend or family car accident or house fire or work incident, then there was always tornado season to look forward to. Where I'm from, we have two of those seasons a year!

It is quite reasonable to see how I could grow up having a considerably skeptical mind toward the positive thinkers out there. I mean, didn't they know that there was some sort of mayhem, destruction, or disaster lurking just around the corner?

To top all that off, I'm also a scientist and an engineer. I've been going to school all my life, it seems, and I turned forty in 2008. I have a doctorate in Optical Science and Engineering, a Masters of Science in Physics, a Masters of Science in Aerospace Engineering, a Masters of Astronomy, and a Bachelors of Science in Electrical Engineering, and I'm a licensed professional engineer in the state of Alabama. So I've been trained not necessarily to be a skeptic, but to at least ask a hell of a lot of questions about, well, just about everything. So when my young daughter asks, "Why is the sky blue?" or "Why is the grass green?" mommy usually smiles and says, "Go ask your father." And after an hour or so of explanation, drawing charts and graphs on the sidewalk with her sidewalk chalk, and showing her models and simulations from Mathcad on the computer, my daughter usually goes back and asks her mother again. One day she'll learn. I hope not.

I'm also a writer. I've written or co-written about a dozen

science fiction novels and two textbooks. Of course as a scientist I've written or cowritten over two dozen scientific papers on various scientific topics ranging from lasers to quantum physics to rockets and spacecraft for interstellar travel. One of the textbooks was a fun exercise in defending the planet if we were ever attacked by aliens. I'm not a UFO-conspiracy person, but I write a lot of action science-fiction novels about alien invasions, so I thought it would be cool to write a text on how we'd survive for real if it really happened. Besides, an invasion from superior forces from outer space is just another bad thing that could be looming around the corner! Don't tell my grandmas; they'll be awfully disappointed if the invasion never comes around the corner.

Of course, we're not being serious: my coauthors and I found the odds of such a thing occuring to be almost impossible to calculate, so don't fret over it.

My other textbook is a college-level textbook on rocket science and engineering. That one took a lot of work—a LOT of work. And writing it only deepened my desire to follow the scientific method in all things in life.

I have worked for over twenty years now for a multitude of Department of Defense research programs, NASA projects, private industry, and even the U.S. Intelligence Community. It has all been a lot of fun, hard work, and has shaped me to think like a problem solver.

Now, I'm not telling you all this to brag or get you to buy my books or anything like that—though feel free to buy my books if you'd like. My background story is given here to bolster the credibility of this book and the ideas within it. I'm a skeptical Southern boy. I'm a scientist. I'm an engineer. I'm a writer. Keep that in mind.

The reason for telling you all this is that I have made a personal discovery—a revelation, if you will—that by controlling your thought process and thinking in the right way, *you* do indeed have the power to influence *your* world through your thoughts!

Before I get into the details of why and how this works, I need to discuss a little more about how I came to this conclusion, and to do that I have to talk a little more about my personal life.

After all, most truly life-altering, Earth-shaking discoveries and revelations are not from so-called "eureka moments," but instead come from a lifetime of work, study, and personal growth. But then again, sometimes those damned eureka moments will sneak up on you and surprise the living hell out of you, too!

My discovery was partially a long-growth effort of screwing up and learning what I'd done wrong, then starting all over from the beginning . . . or at least from somewhere way closer to the beginning than I would have liked. But there was also a eureka moment.

In fact, there were three.

In high school I was always doing stuff other than school, because to me school was too darned easy. Little did I realize that my small county public school was not equipping me very well for engineering school at Auburn University. My first few quarters at Auburn, I seriously struggled just to pass where high school had been really easy. Why?

I always assumed it was because I had never truly learned how to work hard enough or study properly. So I knuckled down and really hit the books. But still, for the first couple of years in undergraduate college, I struggled throuh some of my classes. Interestingly enough, the classes that I really wanted to take, like Quantum Physics were considered to be the hardest classes—and those were the ones that I made As in. Here I was, a borderline B-C student, making As in the hardest classes. Well, maybe some of that was due to the fact that I was seventeen, in college with pretty girls, and, dare I say it, beer. But with a little perspective on that part of my life, the beer and ladies were not the real problems.

I wanted to be in those harder physics classes. That is where I thought the real study, the real life, of being a scientist or engineer could be found. Those cutting-edge classes were sexy to me. It sure wasn't in the mundane courses that taught us how to hook electrical wires up to motors or basic radio-communication circuits. Heck, I'd done most of that in my high-school science-fair projects.

No, it was in the details of how the universe worked, what was on the inside of the inside of atoms, and quantum physics and General Relativity described the very fabric of reality, and how the electric and magnetic fields somehow miraculously traversed the nothingness of space, of that fabric, even if there were nothing there, a nonexistent fabric, between the stars. It was how we were all made of the same stuff that the stars were made of. It was alluring, as much so as the beer and in its own way, even more intoxicating.

For a teenager to say that studying these subjects was as alluring as partying is hard to believe, I know. But that is the way I felt, even if I hadn't fully realized it at the time.

So, why did I do better in some classes and not the others? Well, again with perspective, I believe it was my attitude. I really wanted to be there. I really wanted to absorb everything I could about those topics. I had a positive attitude. *Hmmm . . .*

As far as the other classes were concerned, I remember saying things like, "I hate this class" or "the instructor doesn't like me" or "he's just trying to fail us". I had a very negative attitude. But fortunately, I wanted to be a scientist so much so that I trudged my way through the muckiness of my attitude and graduated.

I went to work for the army studying the quantum physics of lasers and optics and attended graduate school, working on my first masters degree in physics. My attitude barely changed, and there was always beer. Then I discovered playing guitar and singing in garage rock 'n roll bands, which was another good reason to blow off my studies. But I managed to finish that first graduate degree, and then something wonderful happened — something that changed me forever . . . *for better or for worse.*

I got married. Over the next few years, I became more focused and finished a doctorate, then continued on to get two more graduate degrees. My day jobs were working out pretty good, and my wife and I seemed to be learning from our life mistakes and maturing into smarter adults. I know that sounds awfully squishy for a scientist and an engineer. But it is true.

Then one day I was reading a science fiction novel. When I finished it and set it down, I made the statement, "This book is horrible, and there was no science in the science-fiction!" My wife looked up from what she was doing with an annoyed expression on her face.

"If you think you can do one better, then get up and write one yourself." Now, most of you probably think she was being sarcastic. But she actually said it matter-of-factly, and then nonchalantly went back to what she was doing, as if it were the most natural and obvious idea in the world.

That simple statement would change my life.

A few minutes passed while I sat there, pondering what my wife had just said. Then I got up and walked into the study. I opened up a blank Word document on our computer and started typing. Six months later, I had my first science-fiction novel. Less than a year later, a big science-fiction publisher bought it and its sequel, which was nothing more than a proposal.

But what's most relevant to our discussion is my wife's positive affirmation, which was far more powerful than my critical statement about somebody else's work. With just one positive statement, my wife started us down the road to changing our lives. Of course, *she* was already heading in the right direction, and, as usual, I was dragging up the rear.

That sequel, the one that was merely a rough proposal, ended up being really important as well. That book came to me in a flash. I knew that the title would be *The Quantum Connection*. I also knew that the underlying theme would be about quantum computing and how, like bits in a quantum computer, our universe is connected through a sort of quantum consciousness. Oh, there were plenty of action sequences, hot chicks, cool aliens, and big explosions, but the underlying big idea of the book was that we are all connected to each other through quantum physics.

This is actually true. We know that at least the known part of the universe that we can see was once so small that it was a teeny-tiny point. All the stuff in the universe we see now—the galaxies, the stars, the black holes, the planets, the space between them, the asteroids, the oceans, the skies, the dirt, the grass, the trees, that rock over there, and all of us—was once a single tiny point all connected as one infinitely dense thingy. In physics classes, we call that thingy the Big Bang Singularity. Through physics, we attempt to write a mathematical equation for this singularity. We don't really know exactly how to write it yet, but we're all thinking and working on it.

But the point is that everything in the universe was once all just one thing. Then, for some reason—call it God, a universal will, happenstance, accident, or whatever suits you—it expanded into the universe we see today about fourteen billion years later.

All the stuff from that singularity point spread out and congealed into all the stuff in the universe we see now: the galaxies, the stars, the black holes, the planets, the space between them, the asteroids, the oceans, the skies, the dirt, the grass, the trees, that rock over there, and all of us. We are all a part of the universe.

To be clear: I don't mean that we all are *in* the same universe; I truly mean that we are all *part* of *our one single* universe, and it all began from one singularity point. Therefore we are all physically connected to and entangled with everything and everyone else that is a part of our universe.

Scientists don't really know the full extent of this connection and entanglement, as we are just now devising experiments to study it, but we know it is there. It is this underlying universal connection—this quantum connection—that my second novel is about.

I started working on this novel back around 2002 or so. I researched and researched the details of this quantum connection to the point where I really understood it. I understood so much that I did some actual experiments with it in my day job. I even invented a quantum data bus for a computer for that novel that I don't recall having seen anywhere else. I really began to understand the concept and the physics of it.

Now I've never been really religious. I will spend a little time discussing this from an historical perspective, but very little time. But suffice it to say, whenever the subject of religion would come up, I'd always point out that I was a scientist, and that I'd believe in God when I met him. Until then, I wouldn't have enough evidence to say if he did or didn't exist. Again, I'm a scientist.

But an odd thing was happening to me as I wrote *The Quantum Connection:* I was beginning to believe that there was some overall quantum connection throughout the universe that connected us all. I was beginning not only to believe in it, but to see it as a scientific reality.

A few years went by and I wrote more books, did more science, worked more, and became a father. Life was good. And then I had two eureka moments—almost within a week of each other—that seemed absolutely connected even though they had nothing to do with each other.

The first was that I had been at a science-fiction convention lecturing on various things. I had one panel on "How to Write Hard Science in Science Fiction," and there was another on "Sex in Space." (That last one was a real big hoot.) Then I was on two panels discussing nanomachines and modern physics. Somehow in both of these we spun off into the very small parts of the brain and why or how things happen there and if it was related to this quantum connection. Then one smart guy in the crowd asked if I had ever heard of some new work by this neurologist and a physicist, a famous one whose name he couldn't recall. Apparently the neurologist and the physicist had come up with some new data suggesting that the brain actually works on a quantum scale and not like a neural network as we had previously believed. I had to admit that I had not. I also had to admit that the thought excited the living hell out of me.

That night from the hotel room I put my Google-fu to the test and quickly found the two men in question and the new work. The two men were Stuart Hameroff, M.D. (turned out he was an anesthesiologist, not a neurologist), and Sir Roger Penrose, Ph.D. Their theory is called Orchestrated Objective Reduction (referred to as Orch OR).

I will get into the details of this theory later on in the book (don't worry, there's no math involved), but the basics are that the brain works like a quantum computer and interacts with stimuli through quantum physics.

In quantum computers we set up a question that might be as simple as asking, "Square?" Then we interact that question with the database around it. Instantly, the question will "cohere" or "reduce" to the thing in the database most like it (i.e., anything with a square shape) and cause all the other possible answers to "decohere." In other words, with quantum computers, like things will stick around and unlike things will vanish.

What's even more interesting is that these things don't have to touch each other to do this, they just have to be quantum connected. In physics speak, we say that their mathematical wave functions are entangled.

But before we get too weighed down in quantum physics, the important eureka moment for me was that our brains could be quantum computers! I immediately started with a flood of ideas about this. I stayed up all night reading papers on the Internet about Orch OR. If Orch OR theory turns out to be right, then *our brains truly are quantum computers!*

That means that every thought we have sets up a "quantum question" in our brain (this question is called a "quantum state," by the way). Then it interferes that state with the perception of the universe our brain has stored in it from our lifelong stimulations (sights, sounds, dreams, feelings, smells, touches, you name it), and the state causes a reality or observation to occur or reduce into being.

For example, imagine seeing a red sports car go down the street. Immediately your brain sets up the visual state of the thing you just saw. It then interacts that visual state with all your memories and instantly you know the visual state is a red sports car. The like memories, "cohere," or stay, and the unlike memories—say of an orange baboon or silver train or white horse—"decohere" (i.e., go away).

One of the other intriguing things about this phenomenon of quantum connectedness is that not only is the action instantaneous it doesn't matter how far apart the states are separated physically. The connection can truly be from one side of the universe to the other, and the action still happens instantly. Seriously. I'm not making this up. This is well-established physics that has been around since the 1920s or so.

So this night was my second eureka moment.

A week or so later, my third eureka moment occurred. I came home from work one evening, and my wife told me that I had to see this episode of *Oprah* that she had recorded on the DVR.

Now, I don't typically watch Oprah Winfrey, although I do see her as the epitome of the American dream. She came from nothing and grew to be one of the most influential women in the world. So I respect her immensely.

I asked my wife what the episode was about and she responded that I'd just have to watch it. And like I did when she told me to go write my own book, I listened.

We sat down on the couch and fast-forwarded the show through the intro to the start of the programming. There were several folks there talking about this new book and DVD coming out called *The Secret*.

Now, once again, I want to remind you that I'm a skeptical Southern boy, a scientist, an engineer, and a writer who isn't very religious, and have never been a big believer in self-help stuff. So I sat there with an eyebrow raised like Spock from the original *Star Trek* series.

> **"We are mass energy. Everything is energy. EVERYTHING."**
> —*The Secret*

Then one of these people kept talking about how the universe is made of energy and that we are all energy and that there was this so-called "Law of Attraction" out there that says that like attracts like. One of them even used the example that it was like a magnet.

That's when I cringed. Magnets don't work that way, and neither does electrical energy. Opposite polarities of magnetic poles and electric charges attract, not like ones. The north pole of a magnet attracts the south pole. A positive electric charge attracts a negative electric charge. It's probably where the old expression "opposites attract" comes from.

Obviously, these people weren't physicists or engineers. But this wasn't the Discovery or History Channel, so I cut them a little slack. I kept listening and thinking that maybe they just don't quite understand how to say what they were trying to say. I sat patiently and gave them a chance. And I am very glad I did!

After watching that episode of Oprah, I put *The Quantum Connection*, the brain being a quantum computer, and this *Secret* together into one big picture in my head. I realized that these *The Secret* people were right about the positive thinking and Law of Attraction philosophy being prominent throughout man's history. Great thinkers and philosophers throughout history have written about the concept, well, for about as long as we have records, and it could very well predate that. I started surfing the Internet and reading more and more about these philosophies. I dragged out my old philosophy textbook from undergraduate school and flipped through it. Maybe, just maybe, these people had observed something but didn't quite understand how to explain it.

> "Always remember, your focus determines your reality."
> Qui-Gon Jinn from *Star Wars: The Phantom Menace*

> "Every thought we think is creating our future."
> —Louise Hay

A few days later my wife recorded another episode of *Oprah*. This one had a lady named Louise Hay as the guest. And again, my wife told me I needed to watch it with her. So I did.

I remember telling my wife after watching the episode that I felt as if I'd been sitting there and listening to Yoda talk for the last hour. If this philosophy that Louise Hay was discussing had any real scientific basis to it, then I was on the verge of my own next physical, mental, emotional, and spiritual revelation.

It was all fitting together. The pieces of the universal puzzle that I'd been studying since I was old enough to read were falling into place before me. My life's work was amalgamating into something greater than each of the pieces of experience and understanding that I had attained.

We bought the DVD of *The Secret*. We watched it several times. I read books by the stars of the show. I read and listened to books by Louise Hay. We bought her DVD, *You Can Heal Your Life*, and watched it. I educated myself on this philosophy. I couldn't sleep for weeks, I was so excited by the promise that modern quantum physics and this ancient life philosophy held for me, for humanity.

I put together a detailed briefing about the human brain, Orchestrated Objective Reduction, Quantum Connectedness, Quantum Computers, and *The Secret* philosophies. I found where leaders of nations, religions, science, medicine, and even generals throughout history had made statements that fit right into what this group of gurus was teaching. I continued to joke with my wife that they all sounded a lot like a roomful of Yodas, Obi-Wan Kenobis, and Qui-Gon Jinns. But it wasn't a joke, or some made-up sci-fi world: this was real. And I really liked what I was hearing.

I've presented the material I put together at several science fiction conventions, at book signings, and in rooms filled with colleagues of mine that are well-respected physicists and engineers. And every single time I've given this lecture, I get nothing but praise, excitement, and enthusiasm. And every single time, I find a new bit of information about a philosophy or theory that fits right into this universal understanding.

For example, at one science-fiction convention there was an author there who was a Wiccan. Now, I know about two things about Wiccans, and one of them is how to spell it (though I had to look that up), and the other is that I think Stevie Nicks is one. Okay, I know nothing about Wiccans. My grandma would have probably assumed that they worshiped the devil and told me to stay away from them.

But I had known this author for a while, respected her opinions, and knew she was smart. We had been friends for a couple of years, but I had no idea of her religious beliefs before that moment.

So she told me that Wiccans get together in their coven (or whatever they call it) and plan what they are going to perform magic on, and then they focus on what they want to happen. She told me about how they have to keep their mental focus and how they must visualize what they wanted in a certain way with a clear mind, or it wouldn't work. And she told me that it didn't matter how far away the coven was from what they wanted to effect. Then she smiled and said, "I'm glad that science is finally catching up with what we Wiccans have always known."

At the time I was like a deer in headlights and had images

of those freaky girls from the movie *The Craft* rather than the lovely Samantha Stephens running through my head. I politely nodded and promised to think about what she had told me. And I kept my promise.

Another time I gave the backyard-barbecue version of my presentation to my neighbors from a lawn chair over a few beers and bottle rockets on Independence Day. One of my neighbors smiled and said, "Sounds a lot like praying to me."

Suddenly what my Wiccan friend had said hit home. I started really thinking about the teachings of every religion, philosophy, metaphysical belief system, and even science-fiction and fantasy ideology that I could come up with. They all seemed to have a basic "ask and ye shall receive" aspect to them.

Every day, I felt more and more excited about what I had discovered. Humanity's great thinkers throughout history had somehow tapped into the quantum connection without knowing why it worked. Many studies have shown that people who pray or meditate are physically, mentally, and emotionally more fit than those who don't.

I was really looking at the universe differently now. I was looking at life differently. I was looking at humanity differently.

Humanity could really be on the edge of understanding a bigger reality. And I don't mean understanding that there is something more, but instead understanding that there is something more and how to tap into it.

And that is why I decided to sit down and write *The Science Behind The Secret*. My life's journey to this date has led me here, and I feel I need to not only tap into this quantum connection, but help others understand it.

I will explore a lot of the topics I've presented in lectures that I've given over the past couple of years, as well as some new ideas from the cutting-edge of quantum physics. While it's true that most people who discuss quantum physics have been studying math and physics for years, there are core concepts and ideas that require no math whatsoever (unless you want to show your work . . .). I will try my best to explain the concepts without the math to best of my abilities. No, wait, Yoda would say, "There is no try. Only do or do not." And I recall one of those *The Secret* folks saying, "Trying is failing with honor." So I will not *try*. I *will* explain this concept with as little math as possible.

I am taking great pains to explain these concepts in a way that nonscientists can understand, because I feel they should be shouted from every rooftop, taught in every school, told to every man, woman, and child on the face of the Earth, and spread across the universe. Because, as I've discovered through a life of study, ***Your* Quantum Brain Gives *you* the Power to Change *your* World!**

> "Imagination is everything.
> It is the preview of life's coming attractions."
> —Albert Einstein

Chapter 2:

The Secret and the Law of Attraction

> "It would seem, Adeimantus, that the direction in which education starts a man will determine his future life. Does not like always attract like?"
> — Plato from *The Republic*

What is *The Secret*? That is the first question we need to discuss before we can dive deeper into the arguments for or against its scientific validity. In essence, *The Secret* is considered to be the so-called Law of Attraction. I say "so-called" here because it isn't really a law at all following the definitions of science.

From a scientific standpoint, a law is something that we have hypothesized and theorized, and more importantly, experimented and tested diligently, and then argued and debated through peer review, often over generations. An example would be Sir Isaac Newton's Law of Universal Gravitation:

Every mass attracts every other mass by a force pointing along the line intersecting both points. The force is directly proportional to the product of the two masses and inversely proportional to the square of the distance between the point masses

Newton also posited the Laws of Motion, of which there are three:

1. **The Law of Inertia**—an object at rest stays at rest and an object in motion stays in motion unless acted on by an outside force.
2. **The Law of Force and Momentum**—force acting on a body is defined in a straight line and is the product of its mass and acceleration.
3. **The Law of Reciprocals** (also known as The Law of Action and Reaction)—for every action there is an opposite but equal reaction.

These laws have been tested through scientific experiments over and over and verified to the point that the global scientific community considers them "laws of nature."

The Law of Attraction, on the other hand, is not such an "empirical" law. It is more of an adage, along the lines of the so-called Murphy's Law, which states that "anything that can go wrong, will go wrong."

Murphy's Law is actually a philosophical maxim that is likely intended to lead you along the path of the Boy Scout motto of "be prepared," because "anything that can go wrong . . . will." Obviously, this is not a law in the scientific sense, because there is plenty of evidence every day of moments when there is potential for things to go wrong and they didn't, don't, and will not. Hence, it has been proven not to be an empirical law of nature.

Now, the Law of Attraction states that "like attracts like." That is it. That is the basis of *The Secret*. Like attracts like.

Like attracts like? Really? When I first heard this, high school science jumped in the way, and I thought about magnets and how "opposites attract." It is a scientific fact that opposite poles of magnets, as well as electric dipoles, attract each other. Not like attracts like.

On the other hand, there are plenty of other situations in science and modern physics where like does indeed attract like. Newton's Law of Universal Gravitation states that every mass attracts every other mass. One chunk of mass is not unlike any other chunk of mass when it really comes down to it. Atoms are just electrons, protons, and neutrons in various arrangements. There is interferometry, chaos theory, fractal mathematics, quantum physics, and many others. But we'll get to that later.

The point here is that *The Secret* is simply the idea that within our universe like attracts like. I say this isn't a scientific law because it hasn't had a general analysis through the scientific method to rigorously prove it like Newton's various laws. However, it is at the least a maxim, or law in the general philosophical sense, like that of Murphy. Whether it is true all the time is inconsequential; what matters is that we explore this adage: What is this maxim of like attracts like? What does it mean?

We'll start with the quote from Plato's *The Republic* at the beginning of this chapter. I find it quite humorous and unsurprising that both true believers of *The Secret* and true skeptics of *The Secret* use this very quote to argue that Plato was either teaching the Law of Attraction or that he was doing no such thing. How can one sentence be so controversial? Either he was or he wasn't, right?

Let's review the quote once more and have some fun with it.

> **"It would seem, Adeimantus, that the direction in which education starts a man will determine his future life. Does not like always attract like?"**

The skeptic would likely argue that all Plato is saying to Adeimantus here is that educated people tend to travel in the same circles and that those educated in a certain field tend to travel in the same circles as others educated in that. Simply put, plastic surgeons go to plastic-surgery conventions, science-fiction writers go to science-fiction conventions, cops hang out with each other, and so on. And if one starts out studying a particular vocation, the students he is likely to meet are studying the same.

When I was working toward my undergraduate degree in electrical engineering, I mostly saw only other electrical engineering students, and quite often, through camaraderie, we would make fun of the other "less difficult" engineering fields. You see, we electrical engineers knew that our field was the most difficult of all. Nonsense, of course, but one has to be a team player. The other students in the other majors, likewise, made fun of us in the same vein. Their jokes usually involved thick glasses and pocket protectors, neither of which have I ever owned.

But then I went to graduate school in physics, then optics, then aerospace engineering, then astronomy, each time associating with a new "team." Does this not shoot a hole in this interpretation of Plato's statement?

Not really. But it certainly doesn't prove the skeptic right, though he'll cling to this interpretation of Plato's statement to Adeimantus.

The philosopher and life coach and teacher of *The Secret* will tell us that this is clear evidence that Plato was implying something more than birds of a feather flocking together, and that indeed he was telling us the heart of the Law of Attraction—some three hundred eighty years BCE!

And the fact of the matter is Plato is one of the most renowned deep thinkers in history. His words are the tools with which he taught his ideas, and no matter how you parse it, Plato is plainly stating the Law of Attraction.

Now, my interpretation of the statement Plato makes here is just that: my interpretation. Your interpretation is yours. Who is to say which interpretation is the correct one without actually being able to ask Plato himself? After all, why does he put it in such a way as to say, "Does not like always attract like," as if it were an accepted maxim of the universe meaning something that everybody understands?

Situations like this remind me of a Benjamin Franklin adage my dad used to always tell me that goes something like, "A man convinced against his will is at another opinion still."

But still, this is the beauty of *The Secret*. The idea behind it is deeper than just "birds of a feather" or "like attracts like." The idea is that you create your own reality. You create your universe around you by how you perceive it, interpret it, and feel it. And if you don't believe me, or Plato, just wait until we get to the science.

> "Size matters not. Look at me.
> Judge me by my size, do you? Hmm? Hmm.
> And well you should not. For my ally is the Force,
> and a powerful ally it is. Life creates it, makes it grow.
> Its energy surrounds us and binds us. Luminous beings
> are we, not this crude matter. You must feel the Force
> around you; here, between you, me, the tree, the rock,
> everywhere, yes. Even between the land and the ship."
> —Yoda
> from *Star Wars Episode V: The Empire Strikes Back*

Now we should dig a little deeper into what *The Secret* is supposed to be. The Law of Attraction is the basis for it, but as I mentioned, there is more to it than just that. The true essence of *The Secret* is that everything in the universe is made of energy. You, your house, this book, the light bulb above your head illuminating the pages, the light itself that is illuminating the pages, grass, trees, dogs, cats, televisions, cars, dirt, and everything else in between, above, below, around, behind, and elsewhere all are made of energy.

In fact, even the very fabric of the universe—the actual space between interstellar bodies—isn't empty. Space itself is made of energetic oscillations known as quantum vacuum energy fluctuations (often referred to in science-fiction as zero point energy, or ZPE).

Energy at its basic level of description can be described as a vibration, an oscillation, or a wave. So if everything in the universe is energy, and energy is a vibration, then, as Spock would say, it is logical to deduce that *you* are a vibration or wave, as is anything else within the universe. If you have a thought, then there is some energy required in setting that thought up, and that thought energy is a vibration that ripples through the spacetime continuum (everywhere) across the universe.

And, as we'll discuss in detail a little later, these oscillations are quantum-physics phenomena, and quantum physics happens instantaneously across the entire universe. Wow!

I know that might have hurt your head a little. It hurt Einstein's, too, so don't feel bad about it.

Following the basic understanding of waves—and the physics and mathematics describing how they interact with each other—we find that waves that are most like each other "positively interfere" with each other, or "cohere," and amplify each other into something bigger, whereas waves unlike each other "negatively interfere," or "decohere," with each other. The decoherence is also sometimes referred to as "destructive interference," meaning that the two waves destroy each other. We see this in water waves, sound waves, ripples on the strings of a guitar, and in all things in nature when observed at the level where the wave nature of things becomes apparent.

> **"Whatever you think you can or can't do, you're right."**
> **—Henry Ford**

The Secret suggests that if you focus your thoughts on a particular thing, say an attractive spouse and a million dollars, then the energy wave of that thought is set up and vibrated across the universe. The energy waves in the universe most like you having a hot spouse and a million bucks are the ones that will positively interfere with those thoughts and lead you down the right path to attain such a goal.

Now, there are esoteric aspects of how you are supposed to think of the things that you want most, and the experts in *The Secret*, or the Law of Attraction gurus (such as Louise Hay) can explain how to do this far better than I.

My rough interpretation is that if you "wish" you had a hot spouse and a boatload of money, then the universe will respond to your "wishing" and will happily cohere to your continuing to "wish" for this thing rather than actually allowing you to have the object of your wishing. So you need to be very careful how you wish—no, how you *think*—about these things.

In fact, you are supposed to visualize the things you want as if you already have them and "feel" the way you would feel if you did. As Yoda would tell Luke Skywalker, "You must unlearn." The way we have been raised to think of things is quite in contradiction with the way we *should* think of them.

> **"Size is nothing to the universe
> (unlimited abundance if that's what you wish).
> We make the rules on size and time"**
> — *The Secret*

> **"Every great teacher who has ever walked the planet has told you that life was meant to be abundant."**
> — James Ray, a teacher featured in *The Secret*

There is another aspect to *The Secret* that needs to be mentioned here, and that is another so-called law or maxim known as the Law of Abundance. The Law of Abundance states two things. The most important of these to me as a scientist is that the universe is big. You say, "Duh! Of course the universe is big. What idiot doesn't understand that?" Well, yeah, but . . .

The point of this first part of the Law of Abundance is that the universe is so very, very, very, very big that there is an extremely abundant amount of energy within it.

From a physics standpoint I really like this. The universe, at least the observable universe, is estimated to have some 9×10^{21} stars in it. That is a whopping nine billion trillion stars (or written out would be 9,000,000,000,000,000,000,000 stars). Here in the South we'd use a highly technical term and call that a "buttload." Actually, there's a slightly more "eloquent" term, but there might be children present, so I'll refrain.

When we use the mass of an average-size star to guesstimate the mass of the observable universe, we come up with some 3×10^{52} kilograms. That's a three with fifty-two zeros after it.

Think of it this way. A gallon of milk has a mass of about 3.785 kilograms. So, if you divide the mass of the observable universe by the mass of a gallon of milk, you end up with: 7,900,000,000,000,000,000,000,000,000,000,000,000,000,000, 000,000,000, or a 79 with fifty zeros after it! That's roughly how many gallons of milk it would take to fill the universe. A "buttload" of milk, for certain (not to mention a buttload of cows). The universe is big. Got it.

I absolutely believe the first axiom of the Law of Abundance.

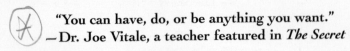

"You can have, do, or be anything you want."
—Dr. Joe Vitale, a teacher featured in *The Secret*

The second axiom of the Law of Abundance is that life is supposed to be lived in abundance. In other words, there is a buttload of energy in the universe, everything in the universe is energy (see previous discussion on the Law of Attraction), and therefore there is no reason for us to not have whatever we want, because there is plenty of it to go around! Why not? Makes sense to me.

The Science Behind The Secret 29

✳ ✳ ✳

Clearly, this is not one of those empirical laws like Newton's, but at the same time the logic is quite impeccable. The skeptic, on the other hand, would say that just because there is an abundance of energy in the universe doesn't mean *you* know how to tap into it.

As Qui-Gon Jinn told Obi-Wan Kenobi in *Star Wars: The Phantom Menace*, "A solution will present itself." I think *The Secret* experts would agree. Again, this comes down to you and how you want to perceive and interpret this.

If you choose to be the skeptic, then that will be your reality. If you choose to perceive that matter and energy and space and time are nothing more than the same entity oscillating in slightly different fashions, then a solution will likely present itself. You create your own reality.

I realize that I keep repeating this, but it is a very important point. Stay tuned and I'll explain it as we go along, and I'll explain the physics behind it to boot. We truly do create our own reality around us.

> **"From the Non-physical, you created you,**
> **and now from the physical, you continue to create,**
> **and we are nothing if we are not Flowers of Energy.**
> **We must have objects of attention, that are ringing**
> **our bells, in order to feel the fullness of who we are,**
> **flowing through us, for the continuation of All-That-Is.**
> **That is what puts the eternalness in eternity."**
> **—Esther Hicks**

Chapter 3:

Where Did *The Secret* and the Law of Attraction Come From?

"The wise ones fashioned speech with their thought, sifting it as grain is sifted through a sieve."
—Buddha

We've discussed what *The Secret* is. Now we should take at least a brief look back through history to where it came from. Most of the Law of Attraction books and Web sites out there talk about this from the New Thought point of view. Few of them, if any, discuss the idea from a truly historical, philosophical, metaphysical, and scientific point of view as well.

Since this book is not just a description of the New Thought ideology or just a scientific analysis of it, but rather a description of my personal journey through the understanding and revelation about the concepts defining *The Secret*, I think we need a new way to look at the history.

There are physics concepts, philosophical concepts, religious concepts, and New Thought concepts that all complement each other and led me to write this book. What follows in this chapter is an overview of the important historical figures who have explored this ideology, and the philosophy and science behind it.

2500-2000 Years Before the Current Era (BCE)

It is really difficult to say where in human history that the philosophy of *The Secret* truly began. Every historian, theologian, metaphysician, and philosopher is likely to know some tidbit of information that can add to this discussion. While this is by no means an exhaustive discussion, it is a description of the major players in developing the philosophy according to my particular journey of discovery.

Sometime before 2000 BCE (that would be over four thousand years ago), a religion or philosophy known as Jainism surfaced in ancient India. Depending on the book or paper you read, this might even be as far back as 3000 BCE. The Jain theology claims that it has always been there and that there is no beginning or end to it, but historically it likely arose during the lifespan of the ancient Indus Valley Civilization in the Indian subcontinent (2600-1900 BCE) created by a person known as Rishabha (see Figure 3.1).

Rishabha apparently taught the members of that ancient civilization agriculture, tending of animals, and cooking—maybe more.

His successors were known as the Tirthankars. The Tirthankars are supposed to be human beings who achieve enlightenment of the perfect knowledge through practicing restraint with respect to the actions of body, speech, and mind—the original Jedi masters. An interesting point about the term "Tirthankars": the word translates as something like "Fordmaker," and they were supposed to lead the way "across the sea of human misery."

The Science Behind The Secret   33

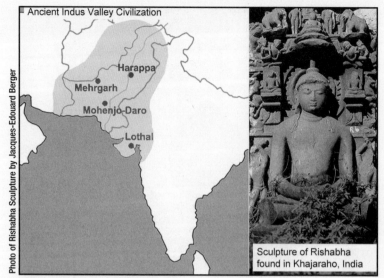

Figure 3.1 Over 4000 years ago, in the Ancient Indus Valley Civilization, Rishabha was perhaps the first teacher of philosophies that would lead to *The Secret*.

Jainism rose in prominence over the course of the reign of the first Tirthankar through to the twenty-fourth. The twenty-fourth Tirthankar, Mahavir Swami, lived between 599-527 BCE (see Figure 3.2). If, as some anthropologists and archaeologists believe, Rishabha lived at some point during the Indus Valley Civilization, then Jainism evolved between 2000 BCE to 527 BCE.

Some of the key philosophies of Jainism are:

- All living beings have a soul.
- Every soul is divine and has (maybe unrealized) infinite knowledge, perception, power, and bliss.
- Every soul is the architect of its own life in this reality and any other afterlife reality.
- Right View, Right Knowledge, and Right Conduct are known as the "triple gems of Jainism."

Figure 3.2 The twenty fourth Tirthankar Mahavir Swami, lived between 599-527 BCE.

The Science Behind The Secret   35

- Control of the senses.
- All souls should strive for liberation from negativity of unenlightened thoughts, speech, and action.
- Strive for Nirvana or Moksha (liberation from the suffering and limitation of worldly existence).

Modern day Jain monks in India, of which there are thousands, practice something known as the Three Guptis (among other things). These are:

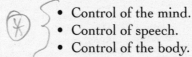

- Control of the mind.
- Control of speech.
- Control of the body.

There is more to Jainism, but this is enough to make the point. We can clearly see how pieces of it are very similar to *The Secret*. To paraphrase, every thing we think, say, and do creates our reality.

Along the same time frame that the Indus Valley Civilization was around, the Old Kingdom period in ancient Egypt was flourishing. It was during this era that the great architectural feats of the Sphinx and the Great Pyramid of Giza were constructed.

More relevant to our topic is that it was during this time that the oldest known religious texts in the world were written—the so-called Pyramid Texts—carved on the walls and sarcophagi of the pyramids at Saqqara sometime between 2400-2300 BCE (again depending on the text or paper one reads this might stretch as far back as 3100 BCE). Perhaps the most pertinent of the statements there is one claiming that *"there are no limits to where the spirit can go."*

This is only a very small line from the Pyramid Texts, (see Figure 3.3) which were mostly concerned with getting the pharaohs into the afterlife, but the one tidbit can be interpreted as though there are no limits to the human spirit and mind.

Figure 3.3 The so-called Pyramid Texts carved on the walls and sarcophagi of the pyramids at Saqqara sometime between 2400-2300 BCE tell us that "there are no limits to where the spirit can go".

Also, spread across the evolution of the ancient Egyptian culture is the idea of Ma'at, which basically translates as "truth." It was the goddess Ma'at that represented the first descriptions of the idea of ethics and the concepts of "truth," "order," and "cosmic balance." Deeper than just being truthful, keeping things in order, and there being a balance of justice in culture, Ma'at represented a universal truth, order, and indeed a balance of all things in the cosmos (see Figure 3.4), which certainly ties in to some of the same philosophies cemented into practice through Jainism.

Figure 3.4 Ma'at represented a universal truth, order, and indeed a balance of all things in the cosmos.

2000-1000 BCE

Over the next thousand years or so, the Egyptians went through various iterations of their religious beliefs and so did the rest of the world. Hinduism evolved during this period, and within that evolution we see prominent the belief of *dharma*. Dharma is the belief in a universal or higher truth as well as an "ultimate reality of the universe". Also from within Hinduism came the ideas of *karma*, *ahimsa*, and *reincarnation*.

Karma is the philosophy that your actions shape your past, present, and future experiences. Many modern metaphysically minded people suggest that karma is actually a force of nature shaping the dharma. There are actually parallel ideas of this in modern quantum physics known as "quantum consciousness" that we will discuss later.

Ahimsa is the concept of doing no harm to other living beings. It has roots all the way back to Jainism. Reincarnation translates to "made flesh again." The idea is that there is some "energy" or "soul" or "spirit" that is the "true higher self" that remains after the flesh dies. At some point this "true higher self" can find its way into a new flesh form and be reborn.

From a physics standpoint, this sounds a lot like an explanation of the Law of Conservation of Energy. This scientific law states that energy can neither be created nor destroyed. If there is some "true higher self" energy then it would seem plausible that it should follow the laws of physics set forth for other forms of energy, as well. But we'll get back to that in later chapters.

Other predominant things within this time frame were the Vedic texts, the prominence of Yoga, the history leading up to Buddhism, and the advent of Abraham. Abraham is probably the most influential figure during the period 2000-1000 BCE. Though his actual time of existing is not quite clear, it does appear to be sometime in this era.

The trials, tribulations, and anecdotes of Abraham are the basis of the morals and philosophies behind the so-called Abrahamic religions of the world. These religions include Christianity, Druze, Islam, Judaism, Rastafarians, and Samaritans. Among the points these religions have in common are a belief in one god (monotheism), an acceptance of prophets, and a belief in some sort of "divine law" that guides the morals and ethics of thier practitioners.

Interestingly, Esther Hicks, a noted Law of Attraction teacher, claims to have gone through a period of meditation for many months following which she could speak with some universal entity or group of entities she refers to as Abraham. Whether or not there is any connection between the name choice of this allegedly channeled entity and historical Abraham is uncertain. But it is interesting nonetheless.

1000-0 BCE

"Do not dwell in the past, do not dream of the future, concentrate the mind on the present moment."
—Buddha

During this time frame is where I believe the basis for *The Secret* began to solidify. And it seems to have happened across the globe in the predominant civilizations of the time almost simultaneously. That in itself is quite astonishing. Or is it? Perhaps the smart guys of the era simply realized there was something to the way they thought, spoke, and acted and what impact that had on the reality around them.

To start with, there was Siddhartha Gautama, who founded Buddhism (563-483 BCE)(see Figure 3.5). At some point in his life it is said that Siddhartha sat beneath a sacred fig tree, now referred to usually as the Bodhi tree, and made the vow to meditate there and not to rise from his position until he had achieved full enlightenment. It is said that Siddhartha

Figure 3.5 Painting of Gautama Buddha's first sermon at the Deer Park.

meditated for forty-nine days and then attained the level of enlightenment he was seeking. From that point on, Siddhartha Gautama was known as the "awakened one" or "enlightened one," or simply, Buddha. He is then said to have understood the nature of all human suffering—ignorance. He set about teaching some fundamentals that became Buddhism:

- Four Noble Truths—suffering is part of existence, the origin of suffering is ignorance, ignorance causes cravings and attachments, attachments and cravings can be overcome, and the Noble Eightfold Path will end the ignorance.
- Noble Eightfold Path—right understanding, right thought, right speech, right action, right livelihood, right effort, right mindfulness, and right concentration.
- Any phenomenon exists because of the existence of all other phenomena in a complex interplay between cause and effect.
- All things are connected through past, present, and future and therefore have no real singular identity (sounds a lot like quantum connectedness, doesn't it?).
- All things are impermanent.
- There is no constant "self."
- All suffering is due to an unclear mind.

These fundamentals sound a heck of lot like major aspects of *The Secret*. The Noble Eightfold Path tells us to watch what we say, do, think, and that we need to be mindful and concentrate in the appropriate way. Very much like *The Secret*.

> **"He who does not understand the Will of God can never be a man of the higher type. He who does not understand the inner law of self-control can never stand firm. He who does not understand the force of words can never know his fellow-men."**
> **—K'ung Fu-tzu (a.k.a. Confucius)**

About the same time far away in China, a different philosophy or religion— or, more aptly, *way* of life—was evolving. Between 551-479 BCE lived a Chinese thinker named Kong Qiu, or K'ung Fu-tzu, later known as Confucius (see Figure 3.6).

Confucius presented himself as a "transmitter who invented nothing." He emphasized to his students and followers that the most important action was that of studying and learning. His approach was not to develop *the* answer and then pass it along, instead, he taught his pupils to think for themselves and reach their own conclusions. His goal was to teach that relentless study of the universe was the path to understanding. Confucius was a firm believer and teacher of the Golden Rule of reciprocity of "do unto others . . . "

"The harder one tries, the more resistance one creates for oneself.

The more one acts in harmony with the universe the more one will achieve, with less effort."
—From the *Tao Te Ching*

Figure 3.6 Confucius was a firm believer and teacher of the Golden Rule or reciprocity of "do unto others . . ."

Figure 3.7 Lao Zi, put together, if not authored, a text known as the *Tao Te Ching* which translates as "the great book of the way to inner strength and virtue"

At the same time as Confucius, one of his contemporaries, Lao Zi, put together, if not authored, a text known as the *Tao Te Ching* (see Figure 3.7). *Tao* means "way" or "path," *te* means "virtue," and *ching* means "classic" or "great book." One possible literal translation of *Tao Te Ching* is the "classical way to virtue." More traditionally, it is "the great book of the way to inner strength and virtue" or something along those lines. From this *way*, the philosophical, almost religious, system of Taoism was born. The basic principles of Taoism are grounded in compassion, moderation, humility, action through inaction, health and longevity, liberty, immortality, spontaneity, and — most relevant to our discussion — a connection to the cosmos.

"Space is the greatest thing, as it contains all things."
—Thales of Miletus

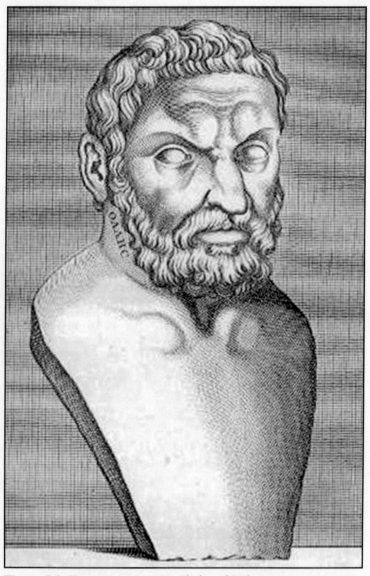

Figure 3.8 From 624-546 BCE Thales of Miletus was considered by even great Ancient Greek philosophers as the "first philosopher in the Greek tradition".

From 624-546 BCE, Thales of Miletus was considered by even great ancient Greek philosophers as the "first philosopher in the Greek tradition" (see Figure 3.8). Many texts on the subject consider Thales as the first true philosopher. Whether that is actually the case or not, he did have several ideas pertinent to our discussion of *The Secret*.

Thales believed in a divinity that perpetuated through all mankind and not just deities. This philosophy led to an idea of a "universal mind." He also spent a good bit of time thinking on geometry and mathematics, leading him to the idea that "space is the greatest thing, as it contains all things." (Of course, he's not referring to outer space, but rather the general notion of empty space.)

Living around the same time as Lao Zi, and following Thales, was Pythagoras of Samos, the Greek scientist, mathematician, and philosopher (see Figure 3.9). (If you took high-school geometry, you certainly would have seen the Pythagorean theorem for calculating the side lengths of triangles.) Pythagoras realized that math was the language which described if not explained the universe. Through his studies he surmised that religion and science must be inseparable. He also adhered to the belief of reincarnation and was the first to propose that thoughts took place in the brain and not the heart.

While it may not be obvious why Pythagoras is key in the evolution of understanding of *The Secret*, it will become clear later. But, suffice it to say that thoughts do occur in our brains, and it is our thoughts that create our reality.

> **"People who take the sun-lit world of the senses to be good and real are living pitifully in a den of evil and ignorance."**
> **—Attributed to Socrates in Plato's *The Republic***

Figure 3.9 Pythagoras of Samos realized that math was the language that described if not explained the universe.

About a hundred years after Lao Zi, Confucius, and Buddha, in ancient Greece arose the classical philosopher Socrates. Socrates formed a method for solving problems known as the "elenchus," which historically is referred to as the Socratic method (see Figure 3.10).

The method is to take a difficult problem and break it down into a series of smaller questions that can be answered, and then continuing along this path until enough questions are answered to lead to an answer of the original problem. This became the basis of the scientific method many years later.

Figure 3.10 Socrates formed a method for solving problems known as the "elenchus" which historically is referred to as the Socratic Method and led to the Scientific Method.

Much like Buddha, Socrates believed that wrongdoing was a consequence of ignorance. He never claimed to be wise but, like Confucius, was a true believer in endlessly studying the path to wisdom. Socrates's main belief was that the best approach for personal growth was to focus on self-development rather than pursuit of material things. It was Socrates who taught Plato, perhaps his greatest pupil (see Figure 3.11).

Plato was a classical Greek philosopher and mathematician and it is through his writings that we learn more about the philosophies of the Socratic school of thought. Plato himself said, "No writing of Plato exists or ever will exist, but those now said to be his are those of a Socrates become beautiful and new." At any rate, the philosophy of Socrates/Plato is quite relevant to our discussion of *The Secret*.

Socrates and Plato were quite intrigued with the concept of reality and the material world. They both seemed to hold contempt for people who could not understand the notion that something could indeed be real even if it did not have a tangible, physical form. Their belief that there were higher insights, higher concepts, and divine inspiration led them to believe that people who couldn't believe in such things could not attain a deeper understanding of reality.

Sounds a lot like Socrates and Plato would argue that accepting the possibility that such things as the Law of Attraction exist might be a path to higher enlightenment. In fact, Plato suggests in his Theory of Forms that the material world is actually a shadow of the real world. I've met a few drunk particle physicists in my day that would say the same thing.

Following Plato was Aristotle and a string of other great thinkers of this era. But clearly the period of time between 1000-0 BCE was a hot bed of thinking, great thinking, and some really great thinking. The basis of *The Secret* was being put together and spread throughout the world to anybody who would listen. And more importantly, the understanding of *The Secret* was still evolving with mankind.

Figure 3.11 Plato was Socrates's greatest pupil and through his writings is perhaps an early teacher of the Law of Attraction

1-1400 Current Era (CE)
"It is done unto you as you believe . . . "
—*Jesus*

√

"Be careful how you think.
Your life is shaped by your thoughts."
—*Proverbs 4:23*
(Good News Bible Translation)

As I readily admit, I've never been a deeply religious person, but clearly, there are numerous Biblical and other religious texts that come from this era that touch upon elements of *The Secret*. Luckily there have also been numerous studies on these, so there is little need for me to dig deeply into them here, other than pointing out that most of the religious texts of this era come from Abrahamic origins.

Noted Middle Ages philosophers include St. Augustine, St. Anselm, Albertus Magnus, and later St. Thomas Aquinas. It is also worth mentioning that a central theme of their work was that science, mathematics, and religion are not mutually exclusive of each other. In our modern times, there often appears to be a divide between religion and the sciences. But the fact of the matter is that both religious philosophers and quantum physicists are asking very similar questions: Is there some greater power at work in the universe? How can we increase our understanding of this? And most important of all, how do we as human beings fit into this universe?

All key questions that *The Secret* seeks to answer.

"There exists no single human being
that does not either potentially
or effectively possess this thing we hold
to constitute happiness."
— Plotinus

Back in Greece, there was a lesser-known philosopher named Plotinus (204-270 CE), who followed in the tradition of Plato (see Figure 3.12). He was instrumental in developing the idea of the "One."

Figure 3.12 Plotinus taught that "there exists no single human being that does not either potentially or effectively possess this thing we hold to constitute happiness." (GNU Free Documentation License image)

The One is the idea that there exists outside of any human experience a supreme and totally transcendent thing or One that cannot be divided and is beyond any concept such as "being." Therefore this One could only exist outside of our reality and must have always been there and must always be there.

Plotinus believed that the One can only be associated with true good and beauty. He also attributes the mere existence of the cosmos as a consequence of the One.

> **"It is vain to do with more what can be done with fewer."**
> **"Plurality should not be posited without necessity."**
> **—William of Occam**

William of Occam (1288-1348 CE) was an English Franciscan friar (see Figure 3.13). He is most known for his logical deduction process known as Occam's razor. His razor basically states that the simplest answer is usually the right one. More specifically, it states that if the solution to a problem can be reached without creating some intangible hypothetical phenomena, then that intangible shouldn't be invoked.

Imagine you're lying in bed in your upstairs bedroom, and you hear scratching at the window. Could it be some strange ghost come to haunt you? Or maybe an animal trying to claw its way in? Occam's razor would suggest a more simple answer, such as a tree branch scratching the pane, or perhaps the cat wants back into the house. Of course, one of the simplest answers is likely the correct one. (But the simple answer isn't always obvious, especially with modern quantum physics.)

More applicable to our topic is Occam's idea of nominalism. In Occam's view, nominalism was the idea that there were no universal truths (such as a God looking in from elsewhere/elsewhen) that existed outside of space and time. Since nobody could prove that "outside of space and time existed," his very own razor would suggest that any universal truth must exist within the realm of the real universe without invoking an intangible hypothetical phenomenon. (I know a lot of string-theory physicists I'd like to explain this idea to.)

In this argument, the intangible is "outside of space and time," whatever that might be. My own personal take on Occam's conclusion here is that he just proved to himself that if there is a god—or a universal quantum consciousness, or an all-encompassing energy source, or whatever—then it must be part of, within, or all of the real universe.

Occam's ideas were actually deeper and more convoluted, but this is my reduction of his philosophy and how it fits with *The Secret*. Basically, through Occam's razor and the notion of nominalism, if we invoke an energy field—or source, or God, or whatever—to define our version of reality, then it must be part of our real universe. We will see soon that the very fabric that makes up our universe fits this description perfectly.

Figure 3.13 Nominalism in William of Occam's final view was the idea that there were no universal truths (such as a God looking in from elsewhere/elsewhen) that existed outside of space and time.

1400 CE - Present Day

"Every completed manifestation, of whatever kind and on whatever scale [is] an unquenchable energy of attraction [that causes objects to] steadily increase in power and definiteness of purpose, until the process of growth is completed and the matured form stands out as an accomplished fact."

—John Ambrose Fleming, (1902)
Electrical engineer and physicist

Throughout this period the growth of metaphysics, philosophy, and religion—as well as science—was immense and far too detailed to discuss in depth here. There were key points such as Sir Isaac Newton's development of calculus (with the unwanted help of Christiaan Huygens and others), the laws of motion, gravity, light, and especially his conception of the universe based on natural law.

It was Newton's belief that there were basic laws of nature, and these led to the modern day ideology of Enlightenment. Great thinkers like Locke and Voltaire applied Newton's ideology to politics and human-rights advocacy.

And in fact, Newton's Law of Universal Gravitation was the first real "Law of Attraction" to gain acceptance as natural law, i.e., scientifically sound. It was about the attraction of one mass to another by means of gravity. And while Einstein showed that this natural law was flawed, it remains a scientifically reliable model on how bodies of mass behave on a large scale (soon we'll be discussing this in greater detail).

It wasn't until the 1800s that people actually started talking about what we now call the Law of Attraction.

Perhaps the first mention of such a concept in the modern era was in a book on esoteric mysteries written by Helena Blavatsky, in 1877 (see Figure 3.14). Blatavsky actually uses the term "Law of Attraction" in her book. Later, in 1902, another writer, James Allen, in the book *As a Man Thinketh*, discusses the idea but doesn't actually call it the Law of Attraction.

Figure 3.14 Helena Blavatsky used the term "Law of Attraction" in a book in 1877.

> **"All matter originates and exists
> only by virtue of a force . . .
> We must assume behind this force the existence
> of a conscious and intelligent Mind.
> This Mind is the matrix of all matter."**
> **—Max Planck**
> **(Nobel Prize in Physics 1918)**

It was also during the 1800s to early 1900s that the era of modern physics sparked into being. John Dalton developed the first theory of the atom around 1803. J. J. Thomson discovered the electron (1897). Ernest Rutherford developed the nuclear theory of the atom.

In 1899, the German physicist Max Planck (Figure 3.15) discovered a fundamental constant of the universe (now called Planck's constant) and realized some very core aspects about light and atoms and how they interacted with steps of energy rather than continuous flow of energy. This was the foundation of the quantum theory of physics.

And a few years later, Planck—along with Niels Bohr, Werner Heisenberg, Erwin Schrödinger, Louis de Broglie, Paul Dirac, and Albert Einstein—ushered in the era of modern physics. We'll discuss their contributions more in the next couple of chapters. But before we postpone the physics, I would like to point out the imput of Louis de Broglie.

Louis de Broglie (see Figure 3.16) showed that any moving particle has an associated wave nature that is based on its momentum (mass multiplied by velocity). He introduced the de Broglie wavelength and showed that it was proportional to the new constant that Max Planck had discovered. So, here we have a new fundamental constant of space and time that enables us to calculate the actual energy-wave properties of solid matter. All objects in our universe have this associated de Broglie wavelength.

Figure 3.15 Max Karl Ernst Ludwig Planck discovered a fundamental constant of the universe (now called Planck's constant) and realized some very fundamental aspects about light and atoms that led to quantum physics.

Figure 3.16 Louis de Broglie theorized that all matter could be represented as "matter waves".

"[The mind itself] plants the nucleus which, if allowed to grow undisturbed, will eventually attract to itself all the conditions necessary for its manifestation in outward visible form."
—Thomas Troward

In the very early 1900s, a man named Thomas Troward made the claim that thought causes physical form. This was the central philosophy of the New Thought Movement. And—as stated by Troward—is essentially the Law of Attraction.

> "You are what you think, not what you think you are."
> — Bruce MacLelland
> from *Prosperity Through Thought Force* (1907)

In 1906, the book *Thought Vibration, or the Law of Attraction in the Thought World* was written by William Walker Atkinson. Clearly the turn of the century was sparking major growth and understanding across several genres of study.

The Law of Attraction was becoming more publicized and written about and was being discussed in every New Age arena from positive thinking, mental science, and practical metaphysics to religious science, science of the mind, and New Thought. There were certainly others. In 1910 there was a book called *The Science of Getting Rich*, by Wallace D. Wattles. The book was based on using the Law of Attraction to gain wealth. A similar book, *Think and Grow Rich* by Napoleon Hill, was published in 1937—again using the Law of Attraction to get rich.

In 1976, cancer survivor Louise Hay (Figure 3.17) wrote a pamphlet explaining her belief that all diseases and ailments were/are due to wrong thinking and the wrong state of mind. The pamphlet quickly spread through the New Age community and led her to writing the book *You Can Heal Your Life*, which was a best seller in 1984. Louise Hay has since written many other books, as well as making audio and video versions of these books. I consider her to be a true Yoda of the Law of Attraction.

Figure 3.17 Louise Hay teaches the belief that all diseases and ailments were/are due to wrong thinking and the wrong state of mind.

In 1988 the couple Jerry and Esther Hicks (see Figure 3.18) published *A New Beginning I: Handbook for Joyous Survival*. In 1991, they published *A New Beginning II: A Personal Handbook to Enhance Your Life, Liberty and Pursuit of Happiness*. Following that, over the next decade or so, were several other books, seven of which were published by Louise Hay's company, Hay House. These include:

- *Ask and It Is Given: Learning to Manifest Your Desires* (2005).
- *The Amazing Power of Deliberate Intent* (2005). *Living the Art of Allowing* (2005).
- *The Law of Attraction: The Basics of the Teachings of Abraham* (2006).
- *Sara, Book 1: Sara Learns the Secret about the Law of Attraction* (2007).
- *Sara, Book 2: Solomon's Fine Featherless Friends* (2007).
- *Sara, Book 3: A Talking Owl Is Worth a Thousand Words!* (2008).
- *Money and the Law of Attraction: Learning to Attract Health, Wealth & Happiness* (2008).

From these books, the Hicks duo claim to be passing along the teachings of an all-intelligent entity or group of entities that Esther Hicks learned to communicate with after many days of meditation. Remember Buddha?

This entity or group of entities, Esther Hicks claims, calls themselves Abraham. The philosophy arising from these books is known as the Abraham-Hicks teachings and is basically as follows:

- You are a physical extension of that which is nonphysical.
- You are here in this body because you chose to be here.
- The basis of your life is freedom; the purpose of your life is joy.
- You are a creator; you create with your every thought.
- Anything that you can imagine is yours to be or do or have.

- You are choosing your creations as you are choosing your thoughts.
- The Universe adores you, for it knows your broadest intentions.
- Relax into your natural well-being. All is well.
- You are a creator of thoughtways on your unique path of joy.
- Actions to be taken and money to be exchanged are by-products of your focus on joy.
- You may appropriately depart your body without illness or pain.
- You can not die; you are everlasting life.

Figure 3.18 Jerry and Esther Hicks are widely known for teaching the Law of Attraction and information apparently told to Esther through channeling an energy entity known to her as Abraham.

In 2004 an independent movie called *What the Bleep Do We Know!?* was produced by students of Ramtha's School of Enlightenment. In this movie, many scientists, philosophers, and New Age thinkers proposed that there exists a spiritual connection between quantum physics and consciousness. The film ended up grossing in the millions of dollars but at the same time received criticism from skeptics that it misrepresented the science in the film. There will always be skeptics.

In March of 2006, The Law of Attraction finally hit the big time, as the larger populace of the planet was exposed to it when Rhonda Byrne's movie, and then book, *The Secret* was released. The DVD and its stars made multiple appearances on widely viewed television programs such as *Oprah* and *The Ellen DeGeneres Show*. It was also featured on *Larry King*. The film and book gained so much publicity that it was spoofed on several television shows, including *Saturday Night Live*.

The really exciting and effective aspect of *The Secret* was the use of so many gurus from different walks of life being interviewed in the film. The so-called "teachers" ranged from self-help authors to preachers, investment specialists, physicists, medical doctors, psychologists, best selling authors, and so on. In my mind, this makes the film more effective than the book. However, they both complement each other.

This far along in our history of *The Secret* you've already been exposed to enough philosophy and teachings that you realize what it is. We can also see that it has roots in philosophy, religion, mathematics, science, and metaphysics. The concepts that evolved into *The Secret* were being formed as much, or more, than four thousand years ago!

And we are still developing an understanding of what the phenomenon is, how it might work, and its impact on reality. There are a couple of other historical events that we need to throw in here to make our picture complete, events that are unfamiliar to most folks.

The Science Behind The Secret 63

> "Consciousness . . . is the phenomenon whereby
> the universe's very existence is made known."
> —Sir Roger Penrose from *The Emperor's New Mind:
> Concerning Computers, Minds, and
> the Laws of Physics* (1989)

In 1989, a book by the noted mathematical physicist Sir Roger Penrose was released (see Figure 3.19). The book was called *The Emperor's New Mind: Concerning Computers, Minds, and the Laws of Physics*. In this book, Penrose argues that human consciousness is not based on any type of algorithm or calculation and therefore cannot be modeled by standard digital computer approaches (the model known as a Turing machine). Penrose goes further, hypothesizing that quantum mechanics must play an important role in the basis of human consciousness and mind.

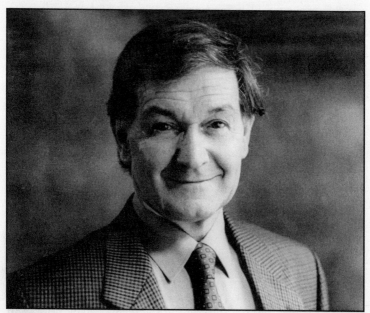

Figure 3.19 Sir Roger Penrose proposed that consciousness must reside within quantum physical phenomena within the brain.

At this point Penrose began collaborating with a fellow named Stuart Hameroff (see Figure 3.20). Hameroff is an anesthesiologist and a professor at the University of Arizona known for his enthusiasm toward the study of human consciousness. Hameroff developed a theory about a part of the living brain cell known as microtubules, which he hypothesized might hold within them some computational capability—potentially quantum computational capability.

The little microtubules are very complicated buggers, and modern science didn't really understand why they needed to be so complicated. Hameroff had read Penrose's book and contacted him regarding his own theories of how the brain works and deals with anesthesia. After all, Hameroff is an anesthesiologist.

The two men met in 1992 and developed a theory based on the microtubules leading to a new understanding in quantum physics and consciousness known as Orchestrated Objective Reduction (Orch OR for short).

In 1994, Penrose published his *Shadows of the Mind*, which describes in some agonizing detail his and Hameroff's theory. Another great place to start learning about this theory is at Hameroff's website, www.quantumconsciousness.org

> **"We are not only observers.**
> **We are participators.**
> **In some strange sense this is a**
> **participatory universe."**
> **—John Archibald Wheeler**

Another proponent that humanity indeed impacts reality through the mind was the late, eminent, and very esteemed American theoretical physicist John Archibald Wheeler. Wheeler is best known for coining the terms "black hole" and "wormhole" in modern gravitational physics.

The Science Behind The Secret 65

Figure 3.20 Stuart Hameroff collaborated with Penrose to develop a quantum theory for consciousness based on microtubule proteins in the brain.

Here's a transcript from a radio interview in 2006:

Wheeler: We are participators in bringing into being not only the near and here but the far away and long ago. We are in this sense participators in bringing about something of the universe in the distant past, and if we have one explanation for what's happening in the distant past, why should we need more?

Radio Host: Many don't agree with John Wheeler, but if he's right, then we and presumably other conscious observers throughout the universe, are the creators—or at least the minds that make the universe manifest.

Wheeler was also one of the developers of the mathematics that describes the entire universe as a single wave phenomenon. This is known as the Wheeler-deWitt equation and also as "the wave function of the Universe."

So, now we have reached a point where our brief history of *The Secret* is up to date. Well mostly. In 2005, a science-fiction novel called *The Quantum Connection* was published. At the heart of that story is an idea based upon the work of Wheeler and Penrose, the notion of a universe connected through quantum physics and a universal quantum consciousness. But that is a shameless plug, so let's move on.

Suffice it to say, *The Secret* is deeply rooted in the history and mind of humanity, which, in turn is in tune with the rest of the universe. We are all part of the singular universal sea of energy. Our individual selves are but subset vibrations on a much larger cosmic sounding board. We just need to learn how to tune our vibrations into harmony with the rest of the universe. We'll see more depth of this in the subsequent chapters.

Chapter 4:

Why Einstein Hated Schrödinger's Cat

> "Reality is merely an illusion,
> although a very persistent one"
> —Albert Einstein

When most people hear about *The Secret* for the first time their response is either "Yeah, right" or "Wow, that is amazing." The "Yeah right" folks are typically skeptical and don't like to believe anything. In many cases, they think that the universe is out to get them or, like my grandma, "If something bad ain't happenin,' then somethin' bad is a fixin' to." These are people who need to see proof beyond a shadow of a doubt that anything outside their comfort zone is real, or they will simply dismiss it.

The "Wow, that is amazing" people, on the other hand, are more eager to try believing in something new. However, they often still have a nagging skepticism deep in the back of their minds—or as the great grandmaster of science-fiction, Robert A. Heinlein, put it, "There ain't no such thing as a free lunch!" (SF geeks like myself say *TANSTAAFL*, pronounced "tan-staff-ul.") A lot of people believe this even if they don't realize it.

And, according to *The Secret* teachers, the negative belief isn't helpful for implementing the Law of Attraction in your favor. In fact, the universe is abundant, and therefore, lunch *is* free. We just have to ask for it. And when the universe offers us the opportunity to eat for free, we have to act quickly and concisely.

What we have to do is prove to ourselves that the universe is abundant, we are all connected with each other and the universe, and our thoughts shape our reality. The way I plan to do this is to teach you the pertinent cosmology and quantum physics as best I can without any mathematics, and show you how it is within this real science of nature that the Law of Attraction operates. Hopefully, this will remove your doubt. No, not hopefully. Let's be positive here. It WILL remove your doubt.

So, let's get started. Like Yoda was so fond of telling Luke Skywalker, "Do or do not. There is no try!" Or, as Dr. Michael Beckwith, one of *The Secret* teachers, put it on that Oprah show I mentioned in Chapter 1, "Trying is failing with honor." Therefore, we WILL learn the right science and we WILL learn that *The Secret* has scientific merit. And, it is real!

So let's get started with our discussion of quantum physics. Don't worry, we'll skip the math. Near the turn of the twentieth century, there were some key discoveries that sparked the advent of quantum physics.

We'll start with the nature of light. At first this may seem completely unrelated to our topic, but hang in there, because the nature of light is the heart of quantum physics. It is through understanding what light is that we will learn a great deal about the rest of the universe and how our very minds cause reality to occur.

The earliest mathematical description of light was developed by Christiaan Huygens in the 1600s. He proposed that light acted as a wave, like ripples on a pond. No matter the source of light, it would propagate out from the source in a sphere surrounding that source in all directions. This he called a spherical wavefront. For a two-dimensional representation of this you can view Figure 4.1, or you could just drop a rock in a pond and watch the circular wavefront spread out from where the rock impacted the water.

For its time, Huygens's theory worked very well both mathematically and observationally in the physical world. His biggest problem was that he wasn't as politically connected as Sir Isaac Newton, who had a theory called the "corpuscular theory" of light.

Newton believed that light existed as particles or corpuscles that worked along the same physics and mathematics as billiard balls on a billiard table (we call it pool where I'm from). Newton's theory remained dominant, based on politics and consensus, rather than any real scientific reasoning. Newton was "the man" in those days, and nobody messed with the man.

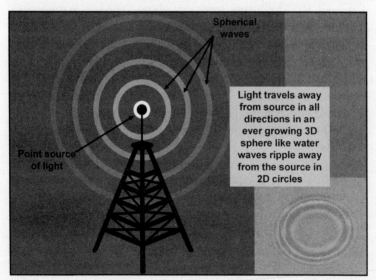

Figure 4.1 A spherical wave propagates outward in a sphere from a central point of origin.

It wasn't until the 1800s when four men conducted experiments and made calculations and developed theories that supported Huygens's wave theory of light so well that the concept of "the electromagnetic wave" became prevalent. The four men were Thomas Young, Augustin-Jean Fresnel, James Clerk Maxwell, and Heinrich Hertz.

Young conducted an experiment that is still studied today in great detail, and many scientists and optical engineers will tell you that they fully understand it, and that we know everything there is to know about it. Well, don't believe them. It ain't true.

In fact, while writing this book, I had a very senior army scientist—whom I respect a good deal—attempt to convince me that we know everything about Young's experiment and that we know what light is. "That has been studied to death," he said, as he tried to convince me to leave the topic alone.

But for all the scientists and engineers that believe we have all the definitive answers about Young's experiment, there are just as many that realize what we are about to discuss here: we have no idea what light is. Light is light. But I'm getting ahead of myself again.

So what was this amazing and elegant experiment that still has scientists studying it over two hundred years later? It was actually very, very simple.

Young took a beam of light and passed it through two very thin slits that were close to each other. The slits were thin as a human hair and were only a few widths of a human hair apart from each other. The material the slits were cut in didn't matter. I've used cardboard, note cards, steel, and aluminum foil. I usually cut the slits with a razor blade. Optics supply companies make fancy precise expensive ones, but cuts in a few cents worth of aluminum foil work just fine.

Young placed his two slits in front of his light source and then put a magnifying glass, or lens, on the other side of the

The Science Behind The Secret   71

slits and focused them down to a spot. (See Figure 4.2). He then observed the light spot at the focus of the lens, where usually there would be one bright focused spot of light, like the bright spot made by a magnifying glass when you focus sunlight onto a fire ant or a piece of paper to start a fire.

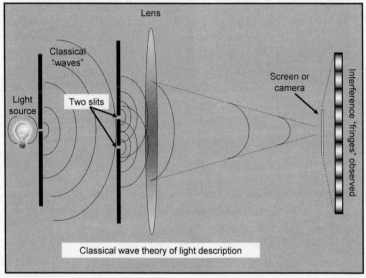

Figure 4.2 Young's double slit experiment shows us the confusing properties of single particles being two places at once!

What Young saw when the two slits were in the way was something unexpected. One might imagine that the result of passing a beam of light through two open slits that were side-by-side would simply be two lines of light.

But Young saw something completely different. He saw several fuzzy lines of light arranged along a straight line horizontal to the slits. Each of the fuzzy lines was surrounded by a dark region. The fuzzy line in the center was the brightest and the ones on either side got sequentially dimmer. (See Figure 4.3).

Why on Earth would there be fuzzy lines with dark bands within them just from passing light through two open slits?

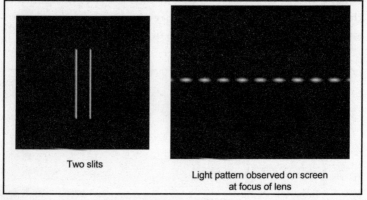

Figure 4.3 Light after passing through Young's two slits.

Upon further analysis, Young realized that if he covered one of the slits, the larger multiple fuzzy lines stayed put but the closer-spaced dark and light bands within went away. He realized that these closer light and dark lines were in fact interference fringes—like the interference patterns that occur when two water waves collide with each other.

You can try these two simple experiments yourself. (Hey, I said there wouldn't be math, but I never said there wouldn't be homework . . .).

Drop two stones in a puddle of water near each other at the same time and watch the waves ripple together into a more complicated wave pattern (see Figure 4.4).

Now set up a bathtub or long, rectangular lasagna pan with an inch or two of water in it and place three blocks in the tub. Arrange the blocks so they look like two slits (see Figure 4.5). The blocks should stick up above water height, and there should be a half inch or so between each of them. Drop a stone in front of the blocks so that a circular wave ripples into them. Watch what happens next.

The Science Behind The Secret 73

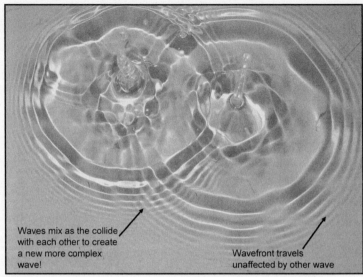

Figure 4.4 Waves ripple together to form destructive and constructive patterns of a new resultant waveform.

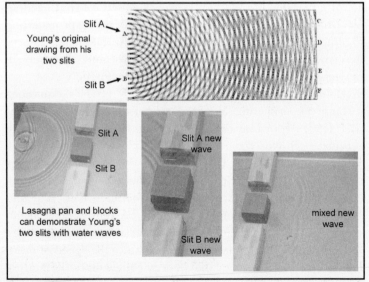

Figure 4.5 A simple bathtub experiment to understand waves and slits.

At each of the openings—our two slits—a new wave is created. The two waves then propagate forward from the slits in half circles. Then the two new waves collide with each other so that on the far end of the tub we have the interference pattern of two slits interacting with the one water wave.

Young realized this and decided then that light must be acting like a wave. Fresnel, a very good mathematician, took the results of Young's double-slit experiment and created a whole new set of mathematical equations describing how light bounced off mirrors, passed through lenses, slits, holes, two slits or two holes, many slits or holes, and many other light-based phenomena. Young and Fresnel had given a lot of credence to Huygens' wave theory of light. And they had created a very powerful model and tool for calculating light phenomena through the mathematics describing reflection, refraction, and diffraction.

Then, in the late 1800s, James Clerk Maxwell created a set of equations known as Maxwell's equations that used an electric-field wave and a magnetic-field wave propagating together to explain light. His new "electromagnetic theory" described light, radio waves, microwaves, and all others in between as these co-propagating electromagnetic waves.

In 1887, Heinrich Hertz verified Maxwell's equations through experiments. Newton's corpuscular theory of light was dead. According to science at the turn of the twentieth century, light was waves.

It was also about this time that quantum physics was being born. Louis de Broglie had explained small matter like electrons, protons, and neutrons as "matter waves" and had shown that any moving matter can be represented as an "energy wave." And Einstein told us that matter and energy were the same thing related by a constant – the speed of light. He told us that $E=mc^2$ or that energy equals mass (of matter) times the speed of light squared. So matter was energy, energy was matter, and they could be represented as waves.

Max Planck had also shown us that the electromagnetic waves must be quantized or come in discrete packets of energy related to his new constant.

Then Einstein took Planck's idea and revived Newton's particle theory of light. He claimed that light was little packets of energy called "quanta." These quanta acted like particles. They could be detected by a certain device known as a phototube. The particles interacted with the material of the phototube as single discrete packets of energy. This was known as the photoelectric effect.

Niels Bohr then generated a model for the atom such that when a single discrete "quantum" of light hit the thing, its outermost electron would get excited and take "quantum leaps" from its normal energy level to an "excited" state. Then, for some reason (which we'll discuss later), the excited electron would get "de-excited" and would emit a single "quantum" of light. These quanta had very discrete and calculable amounts of energy.

Once again a particle theory of light began to dominate, though for some reason the term "quantum" didn't really stick. The quanta became known as "photons." Photons became the particle description of light.

But the problem was that sometimes light acted like a wave—as in Young's double-slit experiment—and sometimes it acted like a particle—as in Bohr's atomic model and the photoelectric effect. And scientists were also learning that matter such as electrons behaved sometimes as waves and other times like particles as well. Physics was getting very confusing.

Then along came Werner Heisenberg. He developed a theory called the Uncertainty Principle, which explained that we could precisely know only the position or the momentum of a particle at any given time but not both (momentum is mass of the object multiplied by the velocity it is moving).

This idea sort of led us back to a wavelike nature of things, since we no longer could think of an electron spinning around an atom's nucleus like the Moon does the Earth. Instead, an electron is whizzing around the nucleus fairly fast, and we have no real idea exactly where it is. The electron became more of an elusive cloudlike thing surrounding the nucleus rather than a single particle. Well, not entirely: that is just one interpretation of the idea. Some would describe the cloud as just the probability of finding the electron in that region around the nucleus at any given time.

Think of it as a fan blade. When the fan is turned on to its highest setting, the blade looks like a solid sheet, and there is no moment where you could identify exactly where the blade is (and I wouldn't recommend sticking your fingers in there to try and find it). But when the fan is turned off, we can see that the fan blades do not make up a solid sheet, and we can easily stick our fingers between the blades.

Electrons around atoms are very similar to this. If the electron's motion were stopped, we'd see a whole bunch of space around the atom. But while it is spinning madly about the atom, the electron generates the perception that it is a solid object surrounding the nucleus of the atom — just like the fan blades.

I know all this sounds strange and maybe even confusing. Believe me, it doesn't just sound that way: it is that way.

To top all that off, Erwin Schrödinger created his quantum wave equation for all things. His equation showed that all things could be described as components of their kinetic and potential energy multiplied by a quantum wave function. Fred Alan Wolf, noted physicist and teacher of *The Secret*, calls this quantum wave function "qwiff" for short.

This quantum wave function, or qwiff, idea first seemed like a good mathematical tool to describe things in the subatomic part of the universe, like electrons, photons, neutrons, and so on. But then Schrödinger realized that there was more to this wave function idea than just the math. As he described in a paper called "Discussions of Probability Relations Between Separated Systems" in 1935:

※ ※ ※

"When two systems of which we know the states by their respective representatives, enter into temporary physical interaction due to known forces between them, and when after a time of mutual influence the systems separate again, then they can no longer be described in the same way as before. ... I would not call that the ONE but rather THE characteristic trait of quantum mechanics, the one that enforces its entire departure from classical ideas of thought."

I believe that Shrödinger's interpretation of his own idea means a lot more than he or humanity would realize for some time to come, but we're getting ahead of ourselves.

What did Schrödinger's statement really mean? Basically, if two objects happen to bump into each other, then the quantum wave functions of those two things become intertwined and "entangled" with each other. The two formerly separate quantum wave functions then become "connected" at the very description of their being.

Bohr went a step further to show that a thing could exist in multiple "states" at the same time until it was observed or measured. The act of observing it caused it to form into only one of the possible states. From Bohr's interpretation, which later became known as the principle of complementarity, the famous Schrödinger's Cat paradox arose. Boy, did Einstein ever dislike this paradox!

Schrödinger proposed to Bohr and the rest of the physics community of the time that if Bohr was right, and it was actually the act of measuring or observing that caused the wave function to collapse into something real, then very weird things could happen.

He proposed that he would take a live cat and put that cat in a box. (That trick in itself should win him a Nobel Prize because nobody can make a cat do something that a cat doesn't want to do.)

Inside this box there would also be a radioactive chunk of

material that had a 50 persent chance of decaying in a certain period of time. If the material did decay, then its resulting radiation would trigger the release of a vial of poisonous gas into the box and that would subsequently kill the cat. If the material didn't decay, then the cat would be well and good and probably wanting the hell out of that box. (Either way, PETA wouldn't like this experiment, but it was just a thought experiment and nobody has ever really tried it, at least not to my knowledge.)

According to Bohr's interpretation of Schrödinger's equation, as long as the box was closed, and we couldn't see in it or hear through it, then the cat must be represented by a quantum wave function for all possible states of which it could exist. There are two: 1) the cat is alive, and 2) the cat is dead. And since there was equal probability of the poison being released, then the cat had equal probability of being alive or dead. Bohr's interpretation of quantum physics says that the cat is actually both dead and alive at the same time until we open the box and observe for ourselves. At that point, our measurement of the cat's state causes one of the states to vanish or collapse or decohere and the other one to appear or cohere.

I know this may seem unlikely or even silly! The cat is either dead or alive, right? That is what the so-called "realist" interpretation of quantum physics would say. But the "orthodox" physicist would say the cat truly is both alive and dead at the same time!

Einstein literally, completely, and absolutely hated this idea! In fact, his exact reaction to it was, "Preposterous!" Einstein's disagreement with Bohr's interpretation of quantum theory led to a split in the field of physics of the time (see Figure 4.6). Many scientists of the time felt as if they had lost the father of physics, meaning Einstein was no longer leading them.

Let's expand on Schrödinger and Bohr a bit more. According to Schrödinger's quantum wave function idea, if we take two particles, each existing in this weird multiple-state configuration (i.e., both existing and not existing), like the cat being dead and alive at the same time, and then let them become "quantum connected" by getting their wave functions entangled, then those two particles will demonstrate what Einstein called "spooky action at a distance."

So in 1935, Einstein and two colleagues, Podolsky and Rosen, published a paper that has become known as the very famous EPR paradox. They originally wrote the paper to prove that quantum mechanics was, if not completely wrong, at least incomplete.

Figure 4.6 Einstein and Bohr debated the meaning of quantum theory and Einstein never felt it was complete.

"Nowadays every Tom, Dick, and Harry, thinks he knows [what a photon is], but he is mistaken."
—Albert Einstein

Here is how the EPR experiment is supposed to go. Consider a neutral pion particle. This is a weird little esoteric particle that shows up in atom-smasher experiments and other cosmic-ray events. (It is also important in tying our understanding of nuclear forces within electromagnetic theory, but that is another story altogether.)

When a neutral pion decays, it creates an electron and an antielectron (called a positron) travelling in different directions, each with opposite spin about its axis. But since electrons and positrons can have two possible spins (either up or down), then each of these particles exists in both states of up and down with equal probability (like the cat being dead and alive at the same time). Not until we measure the spin state of the electron or positron does it actually cohere into a single wave function. Until then, it is a mixed wave function of up and down.

Since these two particles were generated from the quantum event of the pion decay, their quantum wave functions are indeed "connected" from their very beginning, and only one of them can truly be spin up and the other must be spin down.

Einstein, Podolsky, and Rosen went on to show that if we interact with the electron to measure which direction the electron is spinning (up or down) about its axis, then the other particle, the positron, will immediately become the other spin instantly over any distance. Over any distance!

That means that measuring the spin of the electron somehow, faster than the speed of light and instantly, causes the spin of the positron to become the opposite spin. Einstein hoped that the results of this thought experiment would convince people that quantum physics was just too weird to be real and that it was incomplete. He proposed a fix—something called a local hidden variable.

This went unsolved until 1964, when a fellow named John Bell proved that Einstein's fix wouldn't work. And since then, many people have actually demonstrated this spooky action at a distance by setting up the EPR experiment and measuring the results. It not only works with pions decaying into electron/positron pairs but also with photons and other particles.

The experiments show that these particles really do exist in multiple quantum states until they are observed, forcing only one to remain. There are no known cases in macroscopic nature where particles do this, but waves intermingle with each other all the time through interference. (In fact, holograms are just that, interference patterns of multiple waves of light, each in different oscillating form.)

These mixed quantum-state wave functions sound more like, well, wave functions than particles. So, are these waves or particles? How can a particle exist as a mix of different states (spin up or down, dead cat or alive cat)?

Not only does light act like both a particle and a wave, but so does damned near everything else—especially things that are small like electrons. Quantum physics shows us that there is a wave-particle duality simply due to how the object in question is observed.

This brings us back full circle to Young's double-slit experiment and a fellow named Richard Feynman.

> **"It is important to realize that in physics today, we have no knowledge what energy is."**
> **—Richard Feynman,**
> **Nobel Prize winning Physicist**
> **from *The Feynman Lectures on Physics***
> ***Volume I***

I remember at some point reading Richard Feynman's book *QED: The Strange Theory of Light and Matter* and being really flabbergasted at what the "iconic physicist" had said. I was in graduate school and had been studying quantum mechanics. Well, hell, I'd been reading all things quantum mechanics since I was in high school, but it wasn't until grad school that I started really thinking that I understood it.

Of course, later I'd realize that I was full of it and didn't understand it at all. I'm not so sure anybody truly understands quantum physics any more than they understand the tax code, but they are both tools for dealing with certain types of problems—and the question is often one of how to most correctly apply them.

But what was most troubling about Feynman's book was at the beginning of the book, on page fifteen. Here is what he said:

"I want to emphasize that light comes in this form—particles. It is very important to know that light behaves like particles, especially for those of you who have gone to school, where you were probably told something about light behaving like waves. I'm telling you the way it does behave—like particles.

You might say that it's just the photomultiplier that detects light as particles, but no, every instrument that has been designed to be sensitive enough to detect weak light has always ended up discovering the same thing: light is made of particles."

Now this really, really, bothered me back then (around 1994 as I was finishing up my Masters of Science in Physics). I took it on myself to show that you could use wave theory for light and show that anything that light could do could be described by waves.

I wasn't successful, of course, but I also learned that you couldn't do the same with particles, either. One of my mentors and I even wrote a paper or two on the subject (and years later are working on another one with new interpretations of the same questions).

I realized back then that Feynman was either really arrogant or wrong. That really stuck in my gut and tormented me. Feynman was supposed to be one of the greats. I hated to say it, but I just had to: "Surely you're joking, Mr. Feynman!"

Most important of all, over the next few years I learned a very deceptively simple idea: *light is light.*

I realize that sounds redundant and simple and obvious, but it is a far more in-depth statement than it seems. Light *is* light. It isn't a wave or a particle. Waves and particles are, on the other hand, mathematical tools that clever scientists use to describe how light behaves in certain experiments. I had made a personal breakthrough in understanding science not as natural law but as humanity's description of natural law. The two are not quite the same.

How could Feynman have been so arrogant and certain of his knowledge and missed this key aspect of science and humanity?

I reread his book while writing this chapter of the book, and I found this on page eighty-five:

"It's rather interesting to note that electrons looked like particles at first, and their wavish character was later discovered. On the other hand, apart from Newton making a mistake and thinking that light was 'corpuscular,' light looked like waves at first, and its characteristics as a particle were discovered later. In fact, both objects behave somewhat like waves, and somewhat like particles. In order to save ourselves from inventing new words such as 'wavicles,' we have chosen to call these objects 'particles,' but we all know that they obey these rules for drawing and combining arrows [representing complex values of wave functions] that I have been explaining. It appears that *all* the 'particles' in Nature—quarks, gluons, neutrinos, and so forth (which will be discussed in the next lecture)—behave in this quantum mechanical way."

Somehow, back in the mid-1990s, I had managed to forget or miss this paragraph. I had misjudged him. Feynman was actually saying what I had discovered in my personal epiphany about light. And what he was saying is that light is some sort of quantum-mechanical thingy he dubbed a "wavicle" that was something different than a particle or a wave, while sharing characteristics of both.

He also pointed out that all things in nature are made up of things that have these same properties. Everything, in essence, is "wavicles." Just what these wavicles are comes back around to de Broglie's "matter waves" and Schrödinger's wave functions and Fred Alan Wolf's "qwiffs."

Feynman also pointed out that he didn't really care about the weird behavior that light had when it did its uncertain performance with things like two slits. He said, "I am not going to explain how the photons actually decide whether to bounce back or go through; that is not known." He even suggested at times that even asking how or why the photon did these things was a useless question with no meaning.

> **"If we assume that at our core level of being we are all intimately connected in a *unified* field where we are all one, it becomes very easy to understand how we influence one another."**
> **— John Hagelin, Ph.D., teacher of *The Secret***

So what does happen when light hits Young's two slits? Many scientists since have set up experiments so that only one photon is incident on the two slits. I did several similar experiments while writing this book just to remind myself of the little nuances and intricacies of such experiments. Every time I did these single-photon experiments (and anybody else that does them also), I get the same results. If you look for a particle, that is what you find. If you look for a wave, that is what you find. If you look for both that is what you find.

An energy-detection device (Feynman mentioned a photomultiplier tube) always detects single-particles of energy. Always. An interference measurement always gives us a result of waves interfering. And if we watch with an array of single-particle energy-detection devices set up at the far side of Young's double-slit experiment we see single spots hitting the detector array that over time draw (one dot at a time) a wave interference pattern (see Figure 4.7 for results from single photon double-slit experiments).

But if the "light particles" are thrown at the slits one at a time, with several seconds between them, they still map out the interference fringes! How does the photon from ten seconds ago know what the photon ten seconds later had in mind? They had to understand where each other were going in order to map out the right picture! Here is where Feynman would say that physics just describes what we see, so quit trying to understand why things look the way the do and just describe what is there.

I'm not so certain that I believe this interpretation of light (and other wavicles). I do agree that Feynman's statement about what quantum physics tells us is true. It is just a description of what we have observed and not the "why" or the "how."

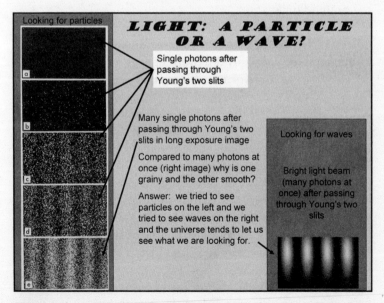

Figure 4.7 Light: A Particle or Wave? What happens when a single photon goes through two slits at once?

That is a very important point to understand about science. Science isn't the "law of nature." Instead, it is humanity's description of that particular law of nature. So, just like Mr. Miyagi explaining to the Karate Kid that "in Okinawa honor very important," we must realize that in science knowing that science is a description and not the law is very important.

Also, note that in the images that when one photon at a time is thrown at the slits, the fringe pattern generated is very grainy and made up of dots. But when the slits are flooded with light, the image is smooth. This is due to the nature of the experiment.

In the single-photon experiment, we were "looking for particles," and that's what we saw, while in the flood-illuminated one, we were "looking for waves," and we saw what we expected. The universe tends to let us find what we are looking for.

Sound familiar?

> **"The more you lose yourself in something bigger than yourself, the more energy you will have."**
> —Norman Vincent Peale

So, what does all this have to do with *The Secret*? At some point in early 2009, I was discussing Young's double-slit experiments with one of my best friends, who's also my wife's sister's husband (does that make him my brother in-law once removed?). Actually, I met him long before I met my wife or his wife because we were study buddies through graduate school. Dr. Pete Erbach and I have spent many nights drinking beer and discussing the esoteric nature of the universe. Well, I was discussing with him the conundrum of Young's double-slit experiment.

If we take a single photon in the vacuum of space away from any other object and describe it with a quantum wave function—a qwiff—then it is quite simple mathematically to write down this qwiff. Then, if we stick two slits in front of the photon, for some reason, that qwiff splits itself into two qwiffs, with a third one that accounts for the other two being entangled.

My question to Pete was, "What is it about that macroscopic chunk of aluminum foil with two slits cut in it that tells the photon to do this?"

At first, Pete responded like Feynman might have with a "that's just the way it is" sort of answer. Then I showed him the math.

First I showed him how Dirac, then Feynman, then one of my mentors, Dr. Frank Duarte (a world-class physicist well known in the quantum optics of lasers world), would each write out the math for the two-slit experiment. The math shows an equation for the photon, then an equation for the slit, then finally an equation for the resultant qwiff on the other side.

Wait a minute: this macroscopic thing with two slits in it is also written as a quantum function? Then, I argued with Pete, the result of Young's double-slit experiment isn't due to the light breaking into multiple wave functions and mixing at all! It must be the light's quantum wave function (or qwiff, that is), interacting with a second qwiff—the one for the electrons, protons, and neutrons that make up the the object with the two slits—and then the outcome is the resultant entangled qwiff.

In other words, the light's qwiff collides with the slits's qwiff. The two qwiffs then mix together, creating a new qwiff that is the next reality observed of the interference pattern measured by our detector array. The math actually looks this way and can be interpreted this way, if we choose to look at it with this perspective—though I'm not certain that anybody else really thinks of this experiment this way.

Pete and I discussed it at length, and thought it worthy of many more discussions over a lot more beer at some time in the near future.

Figures 4.8, Figures 4.9, and 4.10 illustrates the various views of quantum physics as a single photon approaches Young's double-slit experiment.

From the "orthodox view"—as Bohr would have us believe—we see that the photon is truly at all possible locations at the detector plane until it is absorbed by the detector (and therefore detected). All the other possible photon qwiffs vanish, leaving just the one that the detector sees.

Another viewpoint, Feynman's, shows that we don't know anything about where the photon is but the probability for it to have reached a certain location can be determined by following a "probability-density function." There are no multiple quantum states at once (no dead-and-alive cat at the same time, just equal probability that it is one or the other). Once the photon is detected, then we see where it was, and the more photons we use sooner or later the complete probability-density function will be mapped out. Feynman's viewpoint is a bit more pragmatic, that of the "realist."

The final view is more along the lines of what I'm discussing with you here. There is a qwiff for the photon and one for the slits. The two qwiffs interact, creating a third qwiff that is detected by the detector array.

My guess is that none of these are completely correct descriptions of light, and each has its own interesting merits.

But the main point of all this is that from this last point of view or interpretation, we can see that a macroscopic object does indeed interact with microscopic fundamental objects like photons. This suggests that the universe really is all one big bunch of interacting qwiffs intertwined and entangled and connected with each other.

The fact is that something as simple as passing light through two slits in a sheet of aluminum foil shows us that we don't completely understand the universe at either the macroscopic or quantum scale must accept that both are truly pieces of the bigger picture.

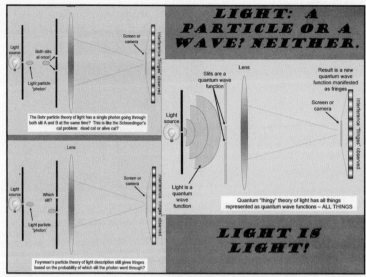

Figure 4.8 Three interpretations of quantum physics and Young's double slit. What happens when a single photon passes through two slits at once?

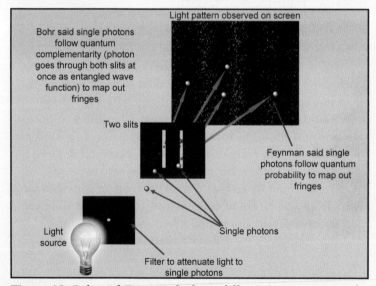

Figure 4.9 Bohr and Feynman had two different interpretations of what happens when a single photon approaches two slits.

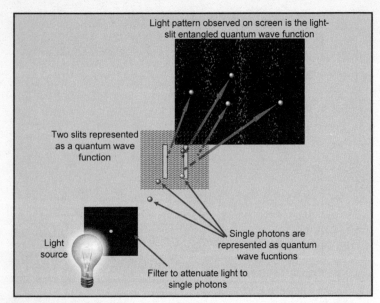

Figure 4.10 Interpretation of quantum physics describing all things as a quantum wave function.

> "A human being is a part of a whole,
> called by us universe, a part limited in time and space.
> He experiences himself, his thoughts and feelings
> as something separated from the rest . . . a kind of optical
> delusion of his consciousness. This delusion is a kind of
> prison for us, restricting us to our personal desires
> and to affection for a few persons nearest to us.
> Our task must be to free ourselves from this prison by
> widening our circle of compassion to embrace all living
> creatures and the whole of nature in its beauty."
> —Albert Einstein

Chapter 5:

There are Computers and then there are Quantum Computers

> "Most explanations portray the brain as a computer,
> with nerve cells ("neurons") and their
> synaptic connections acting as simple switches.
> However computation alone cannot explain
> why we have feelings and awareness, an "inner life.""
> —Stuart Hameroff

Most people in the modern world have been exposed to computers in some form or fashion. Almost everybody from one corner of the Earth to the other has at least heard of the computer. And in the United States and most of the first-world countries everybody has used a computer to surf the Internet. The World Wide Web is as much a part of our daily lives as radios, cars, televisions, cell phones, and toothbrushes are.

Many of us have played video games. Our cars have computer control systems in them. Children's toys have many computers built into them. I've even seen a baseball that has a sensor and a computer built into it in order to calculate the speed at which it was thrown. Not sure that thing can survive being pummeled out of the park with a baseball bat though! The point is . . . computers are everywhere.

So, I don't plan on spending a whole bunch of time explaining to you what a computer is. This is not a computer textbook. However, we need to understand how a normal computer works, what is known as a von Neumann architecture-based computer (i.e., the kind Bill Gates made famous). These typical computers all have a basic hardware design as shown in Figure 5.1. Information is entered through an input device like a keyboard, mouse, microphone, web camera, and so on, which can be processed in a processor (like those made by Intel and AMD and IBM) or stored in the storage component (hard drive, DVD, CD, USB drive, etc.) of the computer. Once processing is done, the processed information is then read through an output device. The user sees this information nowadays on a computer monitor screen, or a printout, or audio output (i.e., sound), etc.

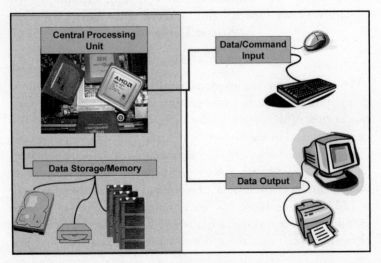

Figure 5.1 Typical computer architecture

The modern computer chip has evolved into an amazing tool. Some of them are small enough to fit in your wristwatch and perform timekeeping, stopwatch functions, calculator functions, and various other things. Some computers are as large as basketball coliseums and consist of millions of computer processors tied together. These are usually called supercomputers and are used for running extremely complex calculations and models and simulations of things like weather patterns. The supercomputers are very expensive and in fact are an ingenious display of mankind's grasp of electronic computing. Most people are more familiar with the desktop and laptop-style PCs. Those may have one or many processors in them, depending on how much money they cost.

But modern computers are limited by the amount of information that can be forced through a wire in the form of an electric current (which is another form of energy). The electric current usually represents a piece of information, and only so much information can be forced through a wire at one given time. If we try to force too much current down a wire, it will melt!

Also, that electrical current traveling through the wire is limited in speed to about 90 to 99 percent of the speed of light. Even if the computer uses fiber optics instead of wires, the speed at which the information can travel is still limited to the speed of light.

As we've already discussed, quantum physics is not limited by the speed of light. Nor is it limited in the amount of information that can be transmitted across wires—as long as two things are quantumly entangled, they can share data instantaneously across any distance. These characteristics of quantum physics make it a good candidate for a very amazing computer, a quantum computer.

Recall our discussion of Schrödinger's cat, where we learned that quantum physics describes all of the possibilities, or possible outcomes, of a quantum event through a representation of those outcomes as quantum wave functions. Each quantum wave function, or qwiff, represents a particular outcome, the cat is either dead or alive. By observing the cat, the actual state of the cat (whether it is alive or dead) coheres to become reality while the other state vanishes.

Imagine if we could make use of this phenomenon for computational reasons.

As I was growing up in the 1980s, there was a toy known as the Rubik's Cube. This puzzle is a six-sided cube, with each side consisting of nine pieces. Each side of the Rubik's cube when solved should be a single color. There are actually 43,252,003,274,489,856,000 or over 43 quintillion possible configurations of the Rubik's cube. I recall playing with the thing for hours and hours and days and days and weeks on end until I finally managed to solve it. Eventually, I trained my brain to solve the Rubik's Cube in about a minute and a half. Some people manage to solve the cube in just over seven seconds. Apparently, those people understood something I didn't.

Some fairly clever mathematicians recently used a supercomputer to prove Rubik's Cube could be solved from any configuration in only twenty-six moves. Normally, a digital computer would simply go through and calculate the solution based upon every possible initial configuration. But even with the world's fastest supercomputer, going through all possible combinations of the Cube, would take many, many years to calculate. Instead, the mathematicians used clever mathematics and tricks to solve half-solved Cubes. And they only looked at 15,000 half-solved Cube combinations. It still took a supercomputer over sixty-three hours to perform this "slimmed down" calculation.

A quantum computer on the other hand could be programmed in such a way that each possible combination of the cube is represented as a quantum wave function. The solved cube would be represented as a quantum wave function and then superimposed with all of the other possible combinations.

The combination wave functions that were solutions would cohere and remain while all of the others would vanish. This would happen at the speed of quantum physics, which is instantaneous. In other words, a true quantum computer could solve instantly a problem that would take today's fastest computer millions of years to solve. We do need to add here that only the simplest of quantum computers have been constructed and demonstrated to date. However, we do understand how they would work when we ever manage to build one.

Wikipedia states:

"A quantum computer is a device for computation that makes direct use of quantum mechanical phenomena, such as superposition and entanglement, to perform operations on data. The basic principle behind quantum computation is that quantum properties can be used to represent data and perform operations on these data."

It also suggests that the types of problems most suited for quantum computers are problems with the following properties:

1. The only way to solve it is to guess answers repeatedly and check them,
2. There are n possible answers to check,
3. Every possible answer takes the same amount of time to check, and
4. There are no clues about which answers might be better: generating possibilities randomly is just as good as checking them in some special order.

And that is precisely what a quantum computer does. These types of problems are like the password-breaking problems seen in cryptology. We see that by allowing possible outcomes of an event—for example, a solution to a Rubik's Cube—to be represented as a quantum wave function, through entanglement and quantum superposition/coherence we can reduce extremely large numbers of possible outcomes to the actual solutions practically instantaneously.

There are an amazing number of potential possible applications for this concept.

I also find it interesting that standard supercomputers take so long to solve puzzles like the Rubik's Cube while many human brains learn to solve even more complex puzzles much more quickly when applied. Is this an argument for the human brain functioning on a quantum-physics level rather than a classical one?

A bit more complicated explanation of this computational capability is to describe the difference between a classical computer bit versus a quantum computer bit. A bit is the basic unit for any type of computer information and is a zero or a one as far as the computer is concerned and may be represented in the hardware as a plus or minus voltage, or no voltage and any kind of voltage (plus or minus). But regardless of which form, in terms of how a classical computer operates, a classic bit is a zero or one.

A quantum bit, also known as a qubit, can be a zero, a one, or an entangled superposition of the two. Therefore a qubit can carry more information than a single classical bit.

A classical computer also follows a particular set of rules and logic that were developed in 1936 by the mathematician Alan Turing. In the 1940s, these rules were refined by another mathematician, John von Neumann, and led to the basic architecture for all modern classical computers.

These mathematical rules use the logic operators "and" and "or" and their inverses to perform operations on only two bits of information at a time. In other words, the classical computer logic will only allow a single bit to operate on another single bit. As with the Rubik's Cube example above, the logic of quantum computers is such that a single bit—that is a single qubit—can operate on a vast number of other qubits simultaneously. Clearly, the quantum computer is far more versatile and powerful than the classical computer. I just wish we could build the darn thing.

We discussed that the bit is represented by a voltage of some sort in the classical computer. How would we build a quantum computer to utilize a qubit?

Figure 5.2 shows a picture of the so-called Loss-DiVincenzo, of double quantum dot, quantum computer chip. The processor was designed by Daniel Loss and David P. DiVencenzo.

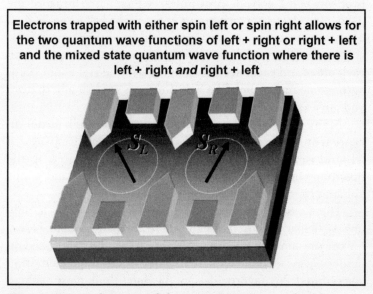

Figure 5.2 A diagram of the Loss-DiVincenzo double quantum dot quantum computer chip

In this picture we see a microcircuit designed in such a way as to trap two free electrons and hold them in place. One characteristic of electrons is that they have spin about an axis much like spinning a basketball on your finger. Some people spin a basketball by tapping at it away from them, and others spin the basketball by tapping it toward them.

Electrons are the same way: They either spin away from you or toward you. Sometimes this is represented as the electron having a spin up or a spin down, and sometimes it is represented as the electrons having a spin left or a spin right. It doesn't really matter which nomenclature you use as long as you pick one and stick to it. Figure 5.2 uses the spin-left and spin-right nomenclature.

The Loss-DiVincenzo qubit chip will trap the electron in either a spin-left or a spin-right position depending on the problem being solved. And unless we actually measure the particular electron pair's status they might be in the "mixed quantum state," meaning they could be either spin-left or spin-right, meaning their wave functions are entangled with each other and each electron is in the left and right spin states at the same time, much like Schrödinger's cat being both dead and alive at the same time.

Many of these chips are strung together in order to represent a particular problem. In other words, strings of left-and right-spinning electrons represent the qubits to the quantum computer in the same way that the plus and minus voltages represent bits in the classical computer.

The Loss-DiVincenzo-based quantum computer is only one particular proposed design. There are many potential ideas for constructing such a device. Many of the ideas are based on modern physics concepts that make their implementation complicated. My favorite design, the one I think most likely to be buildable, is the Loss-DiVencenzo concept.

But what does a quantum computer have to do with *The Secret* and the Law of Attraction? In order to understand *that*, we need to talk about the brain. Because, as we will see in the next chapter, it is quite likely that our brains are actually quantum computers!

Chapter 6:

"Brain and Brain! What is brain?"

> "In my opinion, neurocomputational theories
> fail to explain essential features of consciousness like
> binding, transition from unconscious activities
> to consciousness, non-algorithmic processing and the
> "hard problem" of subjective experience."
> —Stuart Hameroff

"Brain and brain! What is brain?!" That is one of the funniest lines from the original *Star Trek* series. In fact, it is probably my favorite line from the whole series.

It's from the episode where this woman appeared on the bridge of the Enterprise and then disappeared, and when Kirk had a scan of the ship run they found Spock's body but his brain was gone!

When Captain Kirk has a chance to question the woman, in his best interrogator fashion, he asks again and again, "Where is Spock's brain?" She finally replies in a temper tizzy, "Brain and brain! What is brain?"

The funniest part to me has always been the way she said it. All through graduate school—heck, even undergraduate school—I recall many times after endless study sessions and homework sessions somebody in the group looking up and saying that we needed to use our brains to solve the problem, and somebody else responding with that crazy quote from *Star Trek*.

But in terms of our discussion of *The Secret* and quantum mechanics, that is a very valid question for anyone to ask.

"What is brain?"

About fifteen years ago, I was working on a project for the army to create computers that could differentiate vehicles via visual input (i.e., through images from cameras and videos). We attempted to do this using a thing called an artificial neural network. An artificial neural network, or ANN for short, mimics certain processes of how we believed the brain worked.

It was an artificial neural network because we took some studies of real neural networks within the brain and tried to rebuild those using solid-state electronic circuits.

One of the older professors was convinced he could re-create the brain, going down a never-ending path of trying to show that if you add this widget here, and that widget there, and do some more math here, sooner or later he would get to an artificial brain. He was always just one more connection or widget away from success.

Like I said, it was a never-ending path. And he was very unsuccessful as he traveled down that path. We did create some truncated versions of real brainlike neural networks that did decent jobs at identifying, say, a Mustang from a Camaro or an American tank from a Russian tank, which was along the lines of what the army wanted from us. But they were not very efficient at doing so, not truly accurate, and clearly could not perform the function as well as a human brain. If I showed you a Russian tank and an American tank, you might not know which is which, but you would most certainly know that they were different and could see the clear differences. With a minimal amount of training you could easily identify which one was the American tank and which one was the Russian tank. Even a child looking at a Mustang and a Camaro can quickly tell the difference.

Why is it then that our brains work so much better than the artificial neural-network project? It's my belief, and I'm not the only scientist who feels this way, that there is plenty of evidence showing that the brain functions through quantum physics. I've already mentioned in the history section Stuart Hameroff and Sir Roger Penrose and their idea that consciousness must exist through the physics of the quantum scale. And that is what we want to discuss in great detail in this chapter.

There are some skeptics of the concept that the brain works at the quantum level, because, like the old professor that I worked with at the army laboratory, they truly believe that just one more connection, one more circuit, one more processor, one more bit of software, one more final tweak, and their classical model of the brain using artificial neural networks will create the artificial-intelligence brain. I don't really believe in that simply because there is a lot of evidence suggesting otherwise, not to mention my experience on that vehicular-identification project.

One little bit of evidence in particular is how our eyeballs move to study new visual targets and scenes. In a 2004 paper by Behera, Kar, and Elitzur entitled "Quantum Brain: A Recurrent Quantum Neural Network Model to Describe Eye Tracking of Moving Targets," the authors showed that the motion of the eye follows the pattern called *saccadic* motion.

A classical model of the eye would work in such a way that as a new scene, object, or target passed into your field of vision, the eyes would track smoothly from point-to-point, taking in the new scene. The motion would follow a trajectory much like a rocket trajectory and would be a smooth, continuous path.

This trajectory could be predicted by a mathematical function known as a Kalman filter. This is a standard classical method in tracking smooth trajectories.

And if the brain were "classical" rather than "quantum," this is likely how the eyes would track. It turns out that the eyes do not work this way.

Instead, the eyes use this saccadic motion, which is not smooth at all. They focus on a single point and then jump to another point and then jump to another point and continue to jump back and forth and up and down and left and right and over and under and all around the target in order to study it. This paper shows that the classical model of the eye using the Kalman filter represents the eyes' actual motion very poorly. On the other hand, using a quantum physics-based model, the paper demonstrates that the motion of the eye can be represented almost perfectly.

In their conclusion, Behera et al. state quite eloquently:

"Finally, we believe that apart from the computational power derived from quantum computing, quantum learning systems will also provide a potent framework to study the subjective aspects of the nervous system. The challenge to bridge the gap between physical and mental (or objective and subjective) notions of matter may be most successfully met within the framework of quantum learning systems. In this framework, we have proposed a notion of a quantum brain, and a Recurrent Quantum Neural Network has been hypothesized as a first step towards a neural computing model."

This certainly seems to point toward the brain functioning as a quantum system.

There are parts of how the brain works that can't really be calculated, and there's no basic formula that represents those parts. These things are called noncomputable things. When I was growing up, I hated pepperoni pizza. My family loved pepperoni pizza. They always wanted to go eat pizza, and I would end up eating spaghetti. There's no mathematical formula or chemical expression or other classical concept that could explain why I didn't like pepperoni pizza while my family did. We had the same genetics. Of course we were different in that we were different individuals, but we were basically the same organic material with very similar DNA. I didn't have anything that my mom or dad didn't. But I hated pepperoni pizza.

There is no equation to describe why we like some things and hate others. Sir Roger Penrose realized this and suggested that free will is noncomputable and therefore no classical concept can describe that part of the brain's functioning. And, strangely enough, when I got to college I started loving pepperoni pizza and eat it all the time nowadays. Go figure. Noncomputability.

Penrose decided that because the brain is full of these noncomputable aspects, there must be some underlying mechanism that could describe the functioning of the brain and this underlying noncomputability.

There is another (or the same, depending on your interpretation) argument involving a thing called Gödel's theorem, which says kind of the same thing that we are discussing here, but I don't want to get into that as it's a little technical.

Penrose, a very brilliant physicist and mathematician, applied the notion that through Gödel's theorem and this noncomputability of free will, our brains must be working through a means that isn't a simple mathematical operation of one piece of data operating logically on another single piece of data. Each piece of data, or thought, must be functioning through solving many solutions at once without performing the detailed calculations. An experience quickly "coheres" in the brain, as opposed to being slowly calculated. Sound familiar?

Penrose realized this was related to quantum computers and developed a model of the brain that is based on the quantum-physics principle of wave-function superposition. Basically, his idea was that somewhere deep down in the depths of the brain there must be qubit processors very similar to those of the Loss-DiVincenzo quantum processor. But there was a catch.

Modern quantum physics at that time (1994ish) did not require noncomputability. Schrödinger's cat was both dead and alive until either:

1. The box was opened, and the cat was observed or
2. For some reason the cat's wave function got entangled with other quantum sources in the universe and therefore was caused to cohere to whatever state the cat was in.

This version of quantum physics doesn't require consciousness but does require a conscious observation or a bunch of other quantum "noise" to cause the cat to pick a state. This type of quantum physics is called subjective reduction, or SR.

Penrose first decided that there must be a need or reason for the cat's wave function to self-cohere (pick which state it is in) based on something real as opposed to just quantum noise or the act of observation. Perhaps, he concluded, that as the amount of energy in a quantum wave function grows it reaches some threshold that causes this self-coherence.

He then turned to Einstein's General Relativity theory (we'll call it GR for short). GR describes the universe very well on a macroscopic level. Planets, galaxies, stars, and such things down to the size of dust specks can be described well through standard understandings of gravity based on GR. On the other hand, there is no good gravitational theory for things at the quantum scale of atoms, molecules, and subatomic particle like electrons. At this scale quantum physics is needed. There is a gray area between quantum physics and GR where some mix of the two must be the case. This gray area is often referred to as "quantum gravity."

Penrose used a quantum-gravity equation that states that

the amount of time a quantum object can remain in multiple quantum states at once (Schrödinger's cat is dead and alive at the same time) is equal to Planck's constant divided by its total "gravitational self energy" or mass. As an object gets larger the less amount of time it can remain in multiple coherent states at once unless there were some weird "shielding" of normal quantum physics.

From this model, Schrödinger's cat should collapse into a final state (dead or alive) extremely quickly, whether it has been observed or not. This could solve that lingering question about quantum physics. Perhaps. This understanding of quantum physics would require a self-collapse or self-coherence of the quantum wave states and is known as "just objective reduction" ("just OR").

Penrose then went so far as to say that somehow the brain must "orchestrate" this collapse for the noncomputability to take place, and therefore there must be an "orchestrated objective reduction" ("Orch OR") of the quantum states. This was the birth of the theory of Orch OR.

But, how could this orchestrated objective reduction take place in the brain. And where?

Where those qubit processors resided within the brain Penrose wasn't quite certain. And that is where the anesthesiologist Stuart Hameroff came into the picture.

Hameroff was very interested in human consciousness and the mind, so he had been studying the brain for a couple of decades. And he believed that there was something deeper than just the neural networks of the brain, something working at a smaller scale performing much more complicated functions in order to create consciousness. What Hameroff and Penrose have proposed is that the protein called tubulin that resides within many microtubules within a neuron is where this orchestrated objective reduction might take place.

Each of these microtubules is built of billions of tubulin proteins, and each of the tubulins has the capability to trap an electron in a spin-up or spin-down state, just like the

Loss-DiVincenzo quantum processor we discussed in the previous chapter.

The only difference would be that the Loss-DiVincenzo is a computer chip-based technology while the brain's qubits are trapped in organically developed materials (proteins in our brains). There would be billions of the tubulins in each microtubule, tens of thousands of these microtubules in each neuron, and tens of billions of neurons in our brain (see Figure 6.1).

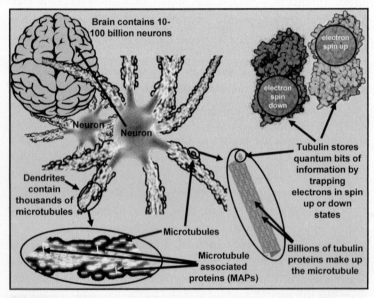

Figure 6.1 The neuron is complex and contains within it the microtubules that may be where quantum interaction takes place in the brain.

"The basic idea is that consciousness involves brain activities coupled to self-organizing ripples in fundamental reality."
—Stuart Hameroff

Penrose and Hameroff had developed a detailed theory explaining the noncomputability of consciousness through quantum physics. Their theory is called Orchestrated Objective Reduction or "Orch OR" for short. In June 2009, Wikipedia described Orch OR as:

". . . a theory of consciousness put forth in the mid-1990s by British physicist Sir Roger Penrose and American anesthesiologist Stuart Hameroff. Whereas most theories assume consciousness emerges from complex computation at the level of synapses among brain neurons, Orch OR involves a specific form of quantum computation which underlies these neuronal synaptic activities. The proposed quantum computations occur in structures inside the brain's neurons called microtubules."

Interestingly enough, this Orch OR theory actually lends itself to the study of the human mind quite well. For a single conscious thought that lasts between a twentieth to a half of a second or so, more than ten billion tubulins must be set up in a quantum state with each other. Then these tubulin states self-cohere to an answer state and a conscious experience occurs. As these Orch OR events occur, a new conscious experience occurs, and the "stream of consciousness" is explained.

Somehow the brain arranges these tubulin states so that conscious thought can be set up to occur. Due to the noisy universe (after all, this is a quantum universe), the conscious moments cannot last much longer than a half second or so. In a single thought of ten billion tubulins, there must be some other mechanism that tunes the thought tubulins, or at least sets up the qubits in the right way within the tubulins as to ask the right question. This is done by other proteins called microtubule associated proteins (MAPs). These MAPs manage this (as shown in Figure 6.1).

Interestingly enough, they also might enable these quantum

oscillations to be communicated to the rest of the brain, and even the outside universe. According to Penrose and Hameroff:

"Microtubule associated proteins (MAPs) attached to certain microtubule tubulin subunits would seem likely to communicate the quantum state to the outside "noisy" random environment, and thereby entangle and collapse it (**SR** or **R**, rather than **OR** or **Orch OR**). We therefore presume that these MAP connections are placed along each [microtubule] at sites which are (temporarily at least) inactive with regard to quantum-coupled conformational changes. We envisage that these connection points are, in effect, "nodes" for [microtubule] quantum oscillations, and thus "orchestrate" the possibilities and probabilities for [microtubule] quantum coherence and subsequent **Orch OR**."

It is, perhaps, this communication to the outside environment that causes the Law of Attraction to take place. These tuned quantum oscillations, these tuned thoughts, are oscillated to the universe whereby the next instant in reality is cohered, creating each individual experience.

In essence, through the physical mechanisms of the quantum theory of Orchestrated Objective Reduction, you really are creating your very own "conscious experience" and your own reality. Your thought oscillations superimpose upon the "noisy" quantum environment of the universe and cohere with something generating your next instant, your next real experience.

If this ain't what *The Secret* Yodas are saying, then I don't know what else it could be!

I realize that this chapter has been very detailed and very confusing. Quantum physics is something that takes years to master. And I'm not certain quantum physics is a topic that *can* be mastered. Then, throwing in some brain organics and philosophy to boot just convolutes the story to heck and gone.

The major point of this chapter is that your brain works like a quantum computer. Each thought we have is due to a set of quantum bits being lined up in just the right way and then are propagated to the rest of the world as a question of "What is my next reality?" That question entangles with other quantum oscillations in the universe, causing one oscillation to occur in your brain. This surviving quantum oscillation (the one that cohered rather than vanishing) is perceived as your next instant of consciousness. Strings of these Orch OR events, these quantum oscillations cohering, make up your "stream of consciousness" and the universe as YOU perceive it.

Chapter 7:

The Qwiff of Your Desire, or Maintaining Your Train of Thought

"Dah dah dah! You stay focus, Daniel-san!"
—Mr. Miyagi from *The Karate Kid*

What we've been discussing throughout this book is quantum physics, various philosophies and religions, and the history of *The Secret*. We've also discussed various pieces of *The Secret*, the Law of Abundance, the Law of Attraction, etc., but what we need to do now is learn how to use this information wisely and to our advantage. After all, what good is an interesting bit of knowledge, a tidbit of trivia, if you don't get to use it in a fun and exciting way, or at least to answer the four-million-dollar question on a game show?

How do we go about using this information?

Fred Alan Wolf is one of *The Secret* teachers from the movie and is also a fairly well-known quantum physicist. In his book *Taking the Quantum Leap* he offered an interesting example to understand wave-particle duality. His example is shown in Figure 7.1: a little drawing that many of us drew when we were in elementary school (and I imagine many of us still doodle on notepads today).

First, you draw one square on the page and then another square slightly offset over it. Than connect the corners with lines making a drawing of a cube. When you look at this drawing what you see is a cube that is either open in the front or open in the back or open on the sides. It depends on how you perceive the drawing.

What you see and what you could say is that this drawing is analogous to light. If you look at this drawing and see that it's open in the front, we will call that a quantum wave function for a front box. If you look at it and see the wave function that shows the box open in the back, we will call that a qwiff for a back box. If you see it open on one side or the other, we will call that a qwiff for side one or side two. Same for the top or bottom.

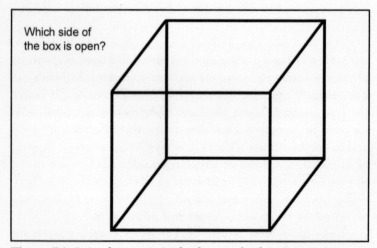

Figure 7.1 Is it a box open in the front or back? Or, is it just a bunch of lines?

This drawing on a simple drawing pad has four potential openings that you'll see one opening at a time. Depending on how good an artist you are, you may only see either the front or the back open and not the sides. The front and back are the only ones we really are concerned with right now, anyway.

In reality what we have on the paper is not a box at all. Instead, it is simply a flat drawing of a bunch of lines arranged in a certain configuration. The truly intriguing thing about this drawing is that it leads us back to the statement that *light is light*.

Recall that light isn't a particle, and light isn't a wave, just like this drawing isn't really what we perceive. Particles and waves are mathematical tools that we use to describe how a natural phenomenon such as light interacts with other things when we do certain experiments. Waves and particles are merely mathematical tools. Light is whatever light is. I can't say this enough: light is not a particle and it is not a wave. If you would like to call light a wavicle, that's up to you. I don't really care. But I would like to say that *light is light* and this box isn't a box—it is merely some lines on a page.

This is a very good analogy and I applaud Dr. Wolf for it. It really does give us a visual example of the particle wave duality of light and quantum physics. And remember, matter has these same wave and particle characteristics as well! Wolf's analogy is also a very good description of the fundamental nature of our reality.

Reality is whatever reality is, and how each individual perceives it will define reality for that individual.

Have you ever seen two kids that have gotten in a fight, and you asked them which one started it? Of course, both of them say the other started it. And, in fact, their individual memories of the fight might support both of them being right, that is, each of them thinks the other is the one that started it. (Assuming one of them isn't just flat-out lying about how it got started.)

So reality exists uniquely in the eye of the beholder. And it is unique for each unique beholder! Especially since, as we've discussed so far, the entire universe is a giant mishmash of energy, an entangled and connected jumble of quantum wave functions. All of these qwiffs—from one end of the universe to the other, from within and without the universe, and everywhere and when in between—make up everything that we see and experience. All of reality starts out in multiple quantum states. All of the possible realities might be existing at once (as per the Schrödinger's cat paradox). Or, according to Orchestrated Objective Reduction, that may not be the case. Reality likely settles into a stable single wave function after a finite interval of time due to Penrose's quantum gravitation concept.

But no matter what the reality is at any given instant, when you have a new thought, you are setting up a new quantum state. With each thought, a new qwiff is generated that begins interacting throughout the universe. Your new thought continues to interact with the universe and with other qwiffs in the universe that are similar to it until a new entangled and common qwiff "coheres" and becomes the next instant's reality. This *is* the essence and the heart of *The Secret*.

"The energy of the mind is the essence of life."
—Aristotle

What you have to do is learn to train your thoughts so that you don't send out qwiffs through the universe that cohere with other qwiffs that in turn generate the next reality that might be undesirable. Don't generate a qwiff that says, "I always get stuck in traffic," unless that is the outcome you desire.

Of course, reality is something much more complex than a single individual perception of it. Like the three-dimensional box that we see on the page that really isn't there, reality is a function of our very own perception and interpretation of it. And if we realize this, then all we have to do is learn how to look at reality and create the outcome that we want. Do we want the box open in the front, or back, or do we just want to see a bunch of lines on the page? It is all these things, and more, but whichever the case may be, we must learn how to make reality be the way we want it to be.

As I mentioned in Chapter 4, if we set up experiments to measure light as a particle, we virtually always measure a particle. If we set up experiments to measure light as a wave we virtually always measure a wave. For some reason, the true essence of light, light's reality, eludes us through direct experiment.

True reality may do the same, but we want to generate our own reality anyway. How can we set up our daily routine and our thoughts so that we measure whatever it is that we're wanting to measure?

Let's start by training our minds to look at the reality of this drawing of the box. Whichever side you see open right now by looking at the image in Figure 7.1, focus on that side and don't let it slip away. Focus and keep the box in one observed state. Now force yourself to see the box open in the front. Focus on it and maintain that the box is open in the front. Do you see it? Now switch that and try to see the box as being open in the back. Focus on it and make it happen. There, see it?

Now try not to see it as a box at all. After all, in reality, it's not a box. It's just a bunch of lines on a page. Don't see a three-dimensional object jump out at you with either an open face or open back, see it as a bunch of lines. You can do it. Make your mind do this. Now see the open side is in the front. Make your mind do it. Open in the back. Just a bunch of lines. Open in the front. Open in the back. Lines. Front. Lines. Back.

Can you do this? This is a very simple exercise in learning to control your mind. Just focus on the open-front box for a couple of minutes. If you can maintain a view of the box that is open in the front for a couple of minutes very readily, then switch to open in the back and maintain that. If you can continue this exercise where you can switch very rapidly from open in the front to open in the back to a bunch of lines to open in the back to a bunch of lines to open in the front and so on at your whimsy, then you are learning to control your mind.

It is a simple exercise, but is very pertinent. As we learned earlier, a single thought only lasts for a few tens of milliseconds. So somehow we must maintain that single thought over and over and over and over in a "train of thought" in order to maintain our focus on what we want. If we maintain this focus, we are sending out the same qwiff into the universe. The more we send this qwiff into the universe, the more likely it is that this wave function will interact and cohere with a qwiff similar to it. And the result will be a reality that we desire.

This is the key to applying *The Secret*—realizing that this is how it works. We create a qwiff with each thought, and we must maintain that thought. This way we manage to continue to send this qwiff of our desire into the universe. In turn, the qwiff interacts with the universe until it meets others in the universe that are similar. Our qwiffs and others out there in the universe collide and create a new one, which becomes the next reality. Hopefully, the new instantaneous reality is the one we wanted in the first place (whether it's a new car, a new baby, a new job, or just something simple, like a tasty lunch, doesn't matter).

In fact, your desired qwiff doesn't have to be materialistic at all. That's something that you have to decide for yourself. What it is you want right here today is the thought you need to be projecting into the universe.

I apologize for being so redundant. This is important; it is the key element you need to understand. With it, you have the basic knowledge to understand how *The Secret* operates and how to ask the universe for the things that you want.

In the 1970s, the psychologist Gaetano Kanizsa generated two diagrams that are similar to Wolf's box analogy. These

Kanizsa figures can be seen in Figure 7.2. When you look at these diagrams, what do you see?

The figure on the left appears to be a triangle with three Pac-Men at the apexes. The one on the right appears to be a square with Pac-Men at the corners. But a second and closer look will reveal to you that there is no triangle and there is no square. The only things in the drawing are the Pac-Men themselves.

For decades now, people have studied these Kanizsa drawings, trying to understand why the brain "sees" the triangle and square even though they are not really there. There has been no real understanding of this until now.

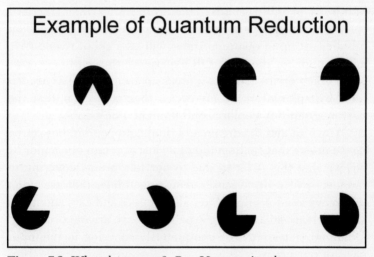

Figure 7.2 What do you see? Pac-Men or triangles and squares? Which ones are really there and why did your brain choose to see what it saw?

Through quantum computational physics (Orch OR) it is likely that the eyes transmit the images as qwiffs in the visual cortex of the brain. Your brain then compares this qwiff with qwiffs of other geometrical shapes through quantum superposition. The quantum oscillations build up until the quantum gravitation threshold is reached and the patterns for the triangle and the square are the ones that cohere. All the qwiffs for all other shapes vanish. At this instant of Orchestrated Objective Reduction, the conscious experience of seeing the triangle and the square occurs. The time elapsed for the experience to occur is a few tens of milliseconds.

Both Wolf's box drawing and the Kanizsa figures display for us a single Orch OR experience. These events happen very quickly—virtually instantaneously. One could surmise from such experiences that thoughts happen fairly quickly as these quantum oscillations settle to a final state within tens to hundreds of milliseconds. Hameroff shows that standard conscious thought patterns that require at least twenty billion tubulins set up in quantum states will take about twenty-five milliseconds for the Orch OR experience to occur.

If you want to maintain a focus on a single goal as needed for the implantation of *The Secret*, then it is clear that the fleeting twenty-five millisecond thought won't do!

As we sit and daydream and think and ponder, how often do we notice that random thoughts just pop into our minds? I believe that this is likely due to our idle brains cohering to random qwiffs from senses detecting things that we don't realize we are detecting. After all, our eyes and ears and noses and touch are still working even though we are daydreaming. Quantum states must be continuously cohering in the input parts of our brains.

Perhaps it is when we decide to focus on a particular subject, feeling, thought, or sensation that something—the MAPs maybe—triggers or sets up the quantum question within our brains. At this point a quantum computation is set up for a reason. When there is no computation set up due to our own free will, then our experience becomes a set of random Orch OR events based on whatever input our brain seizes on for that instant.

So in order to focus our mind on our goals, we must learn to create a steady "train of thought" rather than a string of randomly firing ones. If we think of our goal and then lose our focus, the next thought might be of elephants on the sofa rather than true happiness and inner peace (or whatever your goal is; I'm kinda focusing on a perfect home-brewed beer with just the right frothy head, bittersweet aftertaste, and carbonation just so; well, that and a billion bucks in my bank account!).

> **"Each and every component that makes up your life experience is drawn to you by the powerful Law of Attraction's response to the thoughts you think and the story you tell about your life. Your money and financial assets; your body's state of wellness, clarity, flexibility, size, and shape; your work environment, how you are treated, work satisfaction, and rewards—indeed, the very happiness of your life experience in general—is all happening because of the story that you tell."**
> —Jerry and Esther Hicks
> from *Money and the Law of Attraction*

Seriously focusing your train of thought such that each Orch OR conscious experience is the same for a significant period of time is not as easy as it seems. Try it. It takes practice keeping our minds from being distracted by yesterday's events and random dogs barking and the television in the next room and so on. Practice focusing your mind on your goal for ten or fifteen minutes a day as sort of a form of meditation. Train your mind to control how you set up your quantum computer and keep your train of thought steady.

Why would you do this? It is unclear how well our thoughts might be coupled through the universe and how well they interact. The longer we continue to broadcast the same quantum oscillation, the same qwiff, into the universe then the more likely it is that our projected qwiff will cohere with the desirable quantum events in the universe to lead us to our desired goals. We must stay focused for some period of time. How long of a period of time? Who knows? In *The Secret* DVD, the various Yodas mention focusing for ten or fifteen minutes a day.

> "We are all Vibrational Beings. You're like a receiving mechanism that when you set your tuner to the station, you're going to hear what's playing. Whatever you are focused upon is the way you set your tuner, and when you focus there for as little as 17 seconds, you activate that vibration within you. Once you activate a vibration within you, Law of Attraction begins responding to that vibration, and you're off and running—whether it's something wanted or unwanted"
> —Esther Hicks

Chapter 8:

The Universal Quantum Connection

> "There is a form and level of coherence in the various domains of observation and experience that involves a quasi-instant transmission of information across space and time ... I present evidence that "nonlocal coherence" is widespread in nature, occurring in the microdomain of the quantum, as well as in the mesodomain of life ... "
> —Ervin Laszlo
> from *Science and the Akashic Field: An Integral Theory of Everything*

The universe is abundant in energy and our galaxy alone consists of over 400 billion stars. There are trillions of galaxies. That is a lot of energy. And it is all connected in one quantum universe with a direct tie-in to our quantum brains. Indeed, *your* quantum brain gives *you* the power to change *your* world! Not only your world, but the entire universe is only a thought away. Let's expand on this . . .

Nobody really knows if the entire universe has always been here or not. What we do know is that the observable universe appears to have been here somewhere between 14 and 15 billion years. How the observable universe that may be a part of a bigger universe came about is most likely to have been from some sort of Big Bang. It is this Big Bang that enables us to be connected to everything else in our universe.

Everything in the observable universe that *is* now, was once, 14 billion years ago or so, a single infinitesimally small and infinitely dense point in space and time. In fact, even space and time were part of this single point. This single point—or singularity, if you will—is often called the Big Bang Singularity. We discussed this earlier, but we need to go into a little more detail now.

> **"In some sense man is a microcosm of the universe; therefore what man is, is a clue to the universe. We are enfolded in the universe."**
> **—David Bohm, British Quantum Physicist of the Manhattan Project era**

All that is, all that was, all that ever will be within the observable universe was a single thing that could be described by one really complicated quantum wave function, or qwiff. We'll call this the Big Bang Singularity qwiff. Let's call it the BBS qwiff for short. We have no idea mathematically what the BBS qwiff really looked like, but it is—or was—a qwiff.

As the universe exploded (for whatever reason) and it expanded, the BBS qwiff became the Universal Qwiff. And by Universal Qwiff I don't mean a generic qwiff that describes everything. What I mean is a qwiff that represents the *entire* universe. I realize both of those statements sound almost exactly the same, but think about it a little longer. This isn't a wave function that can represent anything within the universe. We're talking about a single quantum wave function that represents everything within the universe—every single thing. Not just the tangible things like rocks and trees and cars and cats and dogs and girlfriends and children, but also the intangibles like electromagnetic fields and gravity and space and time—everything that exists.

So, as the Universal Qwiff expanded into whatever the universe is today, and again were talking about the observable universe, it got extremely convoluted and complicated because it has to account for all of the individual phenomena that we observe in the universe. And that includes you as a subset of the Universal Qwiff (see Figure 8.1).

Figure 8.1 The universe is a mishmash of all the quantum wave functions mixed together to make up the reality we perceive. Each new thought interacts and joins us through the Universal Quantum Connection.

Your brain is one of these phenomena whether you want to believe it operates like a quantum computer or not, simply because there must be a subset of the Universal Qwiff that describes your brain. Your big toe is a quantum phenomenon whether you want to believe it or not, simply because it is a subset of the larger universal quantum wave function. So, from this discussion it is easy to see that every entity known and even those we have yet to observe are all subsets of the larger Universal Qwiff.

Therefore this *is* a quantum universe. This is a universe that is connected from beginning to end and from within and without. All things in this universe are a piece of a larger thing. A larger quantum thing. All things in this universe are *quantum connected*.

> **"If we do discover a complete theory,
> it should be in time understandable
> in broad principle by everyone.
> Then we shall all, philosophers, scientists,
> and just ordinary people be able to
> take part in the discussion of why we
> and the universe exist."**
> **—Stephen Hawking**

So, how does this play into our discussion on *The Secret*? We've kind of handled this throughout the book, and by now you are probably beginning to surmise this yourself. Reality in this Universal Qwiff is based on the observations that the subset qwiffs (you and me) have within the Universal Qwiff, and how those observations interact with all of the other subset qwiffs within the Universal Qwiff.

Now, realize that when I'm saying "subset qwiff," I really mean you, or me, or your next-door neighbor, because that is what we are. We are a subset of the larger set. You are a subset phenomenon or wave function of the larger quantum phenomenon of the universe. And so is everything else. The larger Universal Qwiff is, mathematically speaking, a linear combination of all other smaller subset qwiffs.

This leads us back full circle to our previous discussions

about every thought you have making your reality. Every thought you have is due to a process of manipulating pieces of your brain to create these thoughts. By altering the small-scale pieces of your brain in order to create a thought, you are manipulating the very fabric of the universe. And then that thought that you have propagates throughout the entire universe instantaneously, because quantum events occur instantaneously across all distance. And that thought qwiff interacts with all of the other qwiffs in the universe.

Does that thought interact with the universe strongly enough to cause a new reality? Or is a single fleeting thought simply quantum noise to the much larger and more energetic universe?

Interestingly enough, *The Secret* movie addresses this (and so does the book, by the way). The noise level of the universe is probably such that single fleeting thoughts are not powerful enough to cohere to a larger qwiff forcing it to become reality.

In other words, to use an example from *The Secret*, if you have the random thought of an elephant in your living room, it doesn't necessarily mean that an elephant is going to appear in your living room. On the other hand, if you continue to focus your thoughts on there being an elephant in your living room, it is possible that the continued perpetuation of this particular qwiff, the thought that an elephant is in your living room, might trigger events to unfold that would deliver said pachyderm to your sofa.

We'll further discuss how to manage your thoughts more particularly in a later chapter. The point of this chapter is for you to realize that you are a subset of the larger whole, that everybody else is a subset of the larger whole, and that we're all *connected*.

This leads to all sorts of possibilities in metaphysical concepts. I've never been a big believer in things like psychic connections, ESP, clairvoyance, déjà vu, psychically connected twins, and other such phenomena. I wouldn't say that I'm a big believer now. However, I would say that I realize there is a real physical mechanism from which these phenomena could occur. This Universal Quantum Connection that we share with all things really does lend itself to some of these potential New Age or metaphysical concepts.

I realize that some of these concepts are not considered mainstream science and are not typically accepted as *real* in mainstream science.

All I am saying here is that the quantum physics we have discussed thus far in this book does potentially offer the mechanism for some of these phenomena to exist, and indeed provides some scientific explanation for these phenomena.

> **"The beginning of knowledge is the discovery of something we do not understand."**
> **— Frank Herbert**

A Ph.D. physicist colleague and friend of mine is a firm believer in ghosts. He is constantly telling me that if I believe in all of this quantum mechanics stuff that I must believe in ghosts. I just tell him that I've never seen one and therefore have no observational experience as a basis for such a belief. I neither believe nor disbelieve in them.

I do have the open mind that there might be a mechanism for them within modern physics, although I certainly don't think they are anything like the ghosts of horror movies. I think they would be more like the energy that is you, your mind, the qwiff that is your driving will once your body can no longer support it.

We know that energy is neither created nor destroyed. Energy merely changes forms. Perhaps somehow within the Universal Qwiff the energy that truly is you blends back into the sea of quantum vacuum energy that makes up everything.

My friend tells me that this sounds like a soul going to heaven.

> "The whole history of science has been the gradual realization that events do not happen in an arbitrary manner, but that they reflect a certain underlying order, which may or may not be divinely inspired."
> —Stephen Hawking

There are plenty of strange phenomena that our new understanding of the universe, the Universal Qwiff and the Universal Quantum Connection, might enable. Let's take twins, for example. In particular, let us consider identical twins that form from the same zygote.

After conception, the zygote splits into what will become two identical twins with very similar genetic codes—almost identical. In fact, the differences in their genetic code are probably only due to malfunctions and mutations and environmental exposure and very tiny differences due to replication. There have been several experiments conducted with twins to determine if there are any unusual mental links—some would say psychic links—between them.

In the book *Science and the Akashic Field* by Ervin Laszlo, several of these experiments are discussed. One particular experiment that comes to mind that is very pertinent to this discussion is one in which twins were taken and put in separate locations. These twins were separated in such a way that there was no contact through any means—including sound, telephones, Internet, smoke signals, radios, Western Union, or any other means of communication. Each twin was then connected to equipment that measured their vital signs, brain patterns, and so on. Then one twin was exposed, without any prior knowledge, to a very loud alarm. In other words, the experimenters scared the hell out of them. In several cases, the vital signs of the other twin simultaneously showed an increase in stress level. How can that be?

It would make sense that at the advance of conception of

the twins there was a strong quantum connection. After all, what really triggers the onset of conception, anyway? We really have only observed the process and not really understood why it happens. Oh, we know what causes it—too much alcohol and short dresses—but we've got no clue why things happen after that part of it.

Perhaps there is an underlying set of quantum events that guide the babies' growth following conception. It would also stand to reason that the zygotes, having both come from the same quantum event of conception, are extremely coupled to one another even after the split. The original zygote before being split was a single subset qwiff. Then it split into two subset qwiffs co-located in the womb together. It would also follow logic that twins should be more closely quantum connected and coupled to each other than, say, two strangers or a person and an oak tree.

In a paper Hameroff published in the journal *Biosystems* he suggests that cell mitosis, the process in which a cell replicates, might be controlled through the mircrotubules and the quantum entanglement between them.

Although the gist of this paper is that during the splitting process errors might occur that could be the cause of cancer, the relevant part to our discussion here is that the splitting of and replication of cells might be controlled via the microtubules and the same type of quantum connection that takes place in the brain. In fact, the microtubules are in all the cytoskeletons of all the cells, which therefore suggests a quantum connection throughout the body.

From the perspective of identical twins, the ramifications are quite intriguing. The initial zygote's splitting and growth might be controlled through the quantum connectivity of the microtubules. The twins should be, at least at some level, connected through the quantum wave function of the original cells from which they were born.

Are experiments like these examples demonstrating quantum connection at a deeper level? Are experiments like

The Science Behind The Secret 129

these merely happenstance, or are they proof of the Universal Quantum Connection? We know mathematically that everything is connected through the Universal Qwiff. Then why shouldn't it make perfect sense that closely coupled subsets of the Universal Qwiff should be able to manipulate one another? Their qwiffs are connected and entangled with each other, and therefore Einstein's "spooky action at a distance" should occur between them.

There is another aspect of the Universal Quantum Connection that I would like to discuss from a personal anecdote standpoint. I know personal anecdotes are not evidence and they don't mean anything as far as proof is concerned—at least not scientific proof. After all, I *am* a scientist, and I typically dismiss personal anecdotes. That is why I can't say I believe in flying saucers. I definitely think the probability of them is substantial, but I'd have to say I don't know if they truly are here, because I've never seen one, nor evidence of them.

All of that said, here's the story. While I was writing this book a family member passed away. It happened while my mother, her sister, and her sister-in-law were on vacation. It was the sister-in-law's mother that passed. And the way we discovered this is the bizarre personal anecdote.

While they were in Las Vegas on vacation, my mother's sister-in-law had a dream that she found her mother in her house, dead. She awoke startled and confused and concerned, so she immediately called home to her mother, who didn't answer the phone. She then called her sister, who lived nearby their mother. My aunt told her to go check on their mother. And, unfortunately, her sister found their mother passed away.

Since I have no reason whatsoever to doubt any of these members of my family, I truly believe that this is the chain of events leading to the discovery my aunt's mother had passed away. To me, these people are reliable sources. To you, it is only anecdotal evidence and proves nothing.

But how many people do you know who have stories like this within their circle of family and friends? I'm certain you know of at least someone who knows someone that has a story like this.

Is it possible that this Universal Quantum Connection that we all share somehow is telling us these things, and for the most part, we just do not understand how to tap into this connection and understand what it is telling us?

> "The problem of synchronicity has puzzled me
> for a long time, ever since the middle twenties,
> when I was investigating the phenomena
> of the collective unconscious
> and kept on coming across connections
> which I simply could not explain as
> chance groupings or 'runs.'
> What I found were 'coincidences' which were
> connected so meaningfully that their
> 'chance' concurrence would represent
> a degree of improbability that would have to be
> expressed by an astronomical figure."
> — Carl Jung,
> Collected Works vol. 8

I've debated with myself over and over about whether I should include one other bit of personal anecdotal evidence. This particular story is very personal and life-changing to me. It happened just a few short years before I wrote this book, and it was one of the things that started me down the path of truly believing in the Universal Quantum Connection.

I was driving home from work as I do most days, and the traffic was no more hectic, than usual. My day had been no more hectic than usual — which is fairly hectic in most cases. But my stress level was no higher than in the yellow, and, after all, I'm in pretty damned good physical shape, so I never worry about that kind of thing. I was less than two miles from home, and suddenly I started getting dizzy. My left arm began to tingle and my chest started to pound and ache. I couldn't breathe and my vision was starting to get blurry. I barely managed to pull off to the side of the road and get my Jeep Cherokee to a stop without crashing it.

I sat there for a few moments wondering if I was having an anxiety attack, or worse, a heart attack. Hell, I was only thirty-eight years old. I was not even going to consider that it was a heart attack. I focused on my breathing and tried my best to relax. All the while I fingered at my cell phone. debating if I should call somebody—maybe even 9-1-1.

But after a few minutes, my breathing evened out. The pain in my chest and arm went away. My vision went back to normal. I felt fine. I started the Jeep back up and made it the rest of the way home. I told my wife about it, and she said I looked tired. She assumed I was just overworking myself and that maybe I ought to go get a checkup and that we should go on vacation. I thought both of those were good ideas.

I didn't think any more about my, whatever kind of attack it was, until a few hours later near bedtime, when I called my brother to ask him about something. I don't even recall what it was—something mundane and trivial. His wife answered the phone.

"Did you hear about Lara?" she asked.

"No, what about her?"

Now, before I go any further, I have to talk about my first cousin for just a bit. Lara was two years younger than me, and we grew up together. We did everything together. When we were very young, we went to church with our grandma and grandpa on Sundays together. We played together. We went to theme parks together. We spent the night at each other's houses all the time. We put on plays and concerts together. We went to our first Kiss concert together (and Motley Crue, and Guns' N 'Roses, and Aerosmith, and many more). We were inseparable, and I love her with all my heart. She was as much a friend as a relative.

Although we both did technical fields in college (she got her degree in mechanical engineering), she was more religious than I. And our different viewpoints led us to many fun and spirited debates as we matured.

So . . .

"Did you hear about Lara?" my brother's wife asked me.

"No, what about her?" I assumed she had made some new brilliant patent where she works. She often did that. "I just talked to her last night about the birthday party this weekend." My daughter was turning three, and Lara had a two-and-a-half and a four-year-old daughter. We had a big weekend planned for them.

"They were driving to go to a movie and they hit a deer. It came through the window and hit her in the head . . . "

After further reviewing of the timeline of events with the accident—as best I can tell—at almost the exact instant I had my "attack," paramedics were on the side of the road somewhere south of Nashville trying to get her heart to start beating again. Lara died at that instant. The paramedics brought her back, but her brain had cascaded from the injury. No matter how hard we all prayed and thought of her being well, she passed on a day or so later. It's likely her quantum computer was so badly damaged from the accident that it just couldn't overcome the trauma. Whatever qwiff described her, her brain was too damaged to stay cohered with it.

It shook me mortally. I had a checkup, a complete physical. Nothing was wrong with me. Nothing. Not a damned thing. There was nothing wrong with me, and there was no explanation for my attack. No scientific explanation . . . except, perhaps, a Universal Quantum Connection. Or maybe it was just a sick, bizarre coincidence.

Again, this is an anecdote, not evidence of anything. It is certainly not scientific evidence. It is "spooky" nonetheless. "Spooky" in the way that quantum physics is "spooky."

I can't say that I even believe it myself. After all, coincidences happen, right? Accidents happen.

Right?

I've often been interested in dreams and how people interpret them and the fact that those interpretations are usually based on some psychological theory that people need to reach closure on some things or it was the last thing they had on their mind when they went to bed, etc. and so on and so on. Over the last few years, I have come to a different hypothesis about dreams.

If your mind is at rest and your will is not forcing your quantum computer brain to set up thought qwiffs to interact with the universe (in other words, What am I seeing? Is that a red sports car? Are those my keys? And so on), then what does the quantum computer do while sitting idle?

The rest of the universe is still broadcasting qwiffs, and your brain, your quantum computer, is part of the rest of the universe. So, are dreams possibly an interpretation of your idle quantum computer latching on to random qwiffs that are being broadcast from other subset qwiffs in the universe?

If that were to be the case, this would have ramifications that are amazing. This would suggest that our brains are not only good at transmitting qwiffs into the universe, but equally capable of receiving qwiffs from the universe. That is amazing in that if we learned to quiet our minds and listen, we might be able to "tune in" our minds to the universe. We might be able to learn to receive quantum information signals from anywhere in the Universal Qwiff.

Another interesting aspect of this universal connection that we all share is that the Universal Qwiff includes all of space and all of time because all of those things are manifestations of our universe. Who knows if time and space have the same meaning outside our universe or in any other universe (if other universes exist)?

Does this mean that we are quantum connected across time to all things? That sounds very intriguing when thinking about anecdotes where people claim that they have lived previous lives or where people think they have contacted dead humans. Oh, I'm not trying to give (or take away) scientific credence to paranormal things. I am merely thinking outside of the box and wondering what the Universal Quantum Connection truly means. To quantum physics, space and time are basically the same thing, the same as energy and matter are basically the same thing. They are the fundamental building blocks of the universe—space and time.

Is it plausible then to think of a person one thousand years ago having thoughts and creating qwiffs that propagate instantaneously through all of the universe to interact with qwiffs created by brains of humans of the present day? Again, time is part of the Universal Qwiff. Is there any possibility that those brain patterns, those subset qwiffs from some humans from long ago, could cohere with qwiffs from humans of the present day?

I don't know if this is nonsense or not. But, we're talking about a connection that is at the very fabric of the universe and these things seem to pop up as weird possibilities the same as Schrödinger and his cat did in the 1930s. These might be the new paradoxes of modern physics in the same light as the Einstein-Podolsky-Rosin experiment was in Einstein's day.

I'm not going to belabor the strange but possible weird ideas. I'm not going to beat a dead horse about something that is purely speculation, something I have no idea how to prove experimentally. All I want to do is to get you to think about things that most people normally don't think about.

If we consider that this Universal Qwiff really does explain everything from the Big Bang until now and into the future, all of the universe, everything and everywhen, then we must consider the consequences of such concepts. We must consider strange—and even "spooky"—possibilities.

We must realize that this Universal Quantum Connection really does go deeper, really is intertwined and entangled with every infinitesimal piece of the universe and with every macroscopic piece of the universe and with everything even on a cosmological scale in the universe.

It is indeed a Universal Quantum Connection that is the mechanism through which *The Secret* operates.

Chapter 9:

Use Your Quantum Brain Wisely

"I will see it when I believe it."
—Wayne Dyer

*I*n reading this book to this point, you have been exposed to a lot of strange and wild and crazy and wonderful ideas, theories, concepts, and intriguing paradigms of New Age thought and wacky quantum physics. We discussed how *The Secret* has a mechanism through the Universal Quantum Connection to operate. We've discussed how your brain functions as a quantum computer, sending out the quantum wave functions to interact through this Universal Quantum Connection.

As we've shown, from the Tirthankars of four to five thousand years ago to Louise Hay of present day and to others throughout mankind's history, many have observed the phenomenon of *The Secret*. These New Age thinkers, great philosophers, and, as I think of them, Yodas, have set out guidelines and ideologies for implementing the power of the universe through appropriate thinking. These views fit precisely with Orch OR and our brains being quantum computers. These views fit precisely with our Universal Quantum Connection. We must learn how to "program" our quantum computers properly so that they interact with the quantum universe in such a way that we get the outcome we desire.

This information is powerful to have. When I first put together the overall picture of *The Secret*, Orch OR, and the Universal Quantum Connection, I felt a little like all the characters in the stories about prophets and such where their eyes were opened to the heavens.

I don't mean I felt like a prophet. I mean I felt as if I had just opened a door to a much bigger universe filled with a much more abundant meaning. Perhaps my feeling was sort of like the first time Luke Skywalker used "the Force." As Luke said, "I really did feel something."

From the history, philosophy, and science of *The Secret* discussed throughout this book, it is clear that there is a good bit of information to learn. Think of it like a martial art of the mind. If you are just now learning of *The Secret*, you are a white belt on your first day of class. Prepare to study, work, and focus. Your future goals should have you moving toward a mastery of the information (attaining your black or red belt, depending on the type of martial art).

Your journey from beginner to master will take many turns, and who knows if you will ever reach true mastery. I learn something new, or a new way to look at things, almost every day. Each new piece of information adds to our understanding of *The Secret* and how deep the rabbit hole truly goes. And we never know where the next lesson will come from. Will it be from a Yoga workout, a science-fiction story, an NFL quarterback being interviewed on television, or a syndicated television talk show? You never know when the next lesson and/or opportunity will come. But when it does, be ready. Act. Take a chance and do something. We need to focus our minds and learn how use our quantum broadcasts wisely.

> **"The greatest discovery of my generation is that human beings can alter their lives by altering their attitudes of mind."**
> —*James Allen*

After all, you wouldn't want to hand someone a powerful tool like a skill saw or laser welder without at least giving them some basic instruction on how to use that particular tool. Earlier, we discussed that fleeting thoughts are probably in the noise level of the universe and not strong enough to cohere and cause any problems. If you have the random fleeting thought of the elephant sitting in your living room, as *The Secret* movie gives an example, you don't really have to concern yourself with an elephant appearing in your living room. But it is possible to train your mind in such a way as to focus it on a single train of thought to achieve the outcome you want.

You must also be careful not to have negative thoughts that your upbringing or society might have instilled in you over the years. "Why do bad things always happen to me?" is never an appropriate state of mind. We have to realize when we are having these types of thoughts and instantly push them out and focus on something else that is positive and happy. The old idea of "going to your happy place" really might be a good idea. The way we express our focus and thoughts is extremely important.

> **"I am secure and am nourished by the love of the Universe itself."**
> —Louise Hay

Mother Teresa was often fond of saying that she would not attend an antiwar rally of any sort, but if one was planning on having a pro-peace assembly, she would be glad to appear. This example is very useful in displaying the difference in the way of saying the same thing. An "antiwar" rally has at its center focus war. A "pro-peace" rally has at its center focus peace. See the difference?

According to *The Secret* gurus and people like Louise Hay, the basis to controlling your thoughts and using your quantum brain wisely is to have positive thoughts. Think of only the positive outcomes!

> **"Creation is always happening. Every time an individual has a thought, or a prolonged, chronic way of thinking, they're in the creation process. Something is going to manifest out of those thoughts."**
> **—Michael Beckwith**

And, more important, envision the outcome of your thoughts and desires having *already* occurred. Create the feelings and emotions of having already accomplished your goal.

Don't just sit there and wish for a new bicycle. We don't say, "I wish I had a new bicycle, I wish I had a new bicycle, I wish I had a new bicycle."

The New Thought community would suggest that the universe would respond to your statements and thoughts of "wish I had a new bicycle" by your reality becoming you *wishing* to have a new bicycle. This makes perfect sense from a quantum wave function standpoint. If you're setting up qwiffs that are propagating out to the universe, and your qwiff says, "I wish I had a new bicycle," then that qwiff is likely to cohere with other qwiffs that create the outcome of you wishing you had a new bicycle.

The right way to use your quantum brain is to set up the thought pattern of the reality that you actually have that new bicycle. You must envision yourself riding that new bicycle. You must fantasize how it feels to ride the new bicycle. What you should *not* do is think, "I wish I had that new bicycle." This is important. You simply envision the new bicycle. You focus on the color of the bike, the way it feels when you sit in the saddle, how the handlebars feel when you grip them the very first time. Feel how elated you are to be on the bicycle riding around the world and enjoying it.

That is how you set up a qwiff to get a response from the universe in a positive manner.

> **"It is curious how often you humans manage to obtain that which you do not want."**
> **—Spock in *Star Trek: Errand of Mercy***

We have to watch and unlearn those negative thoughts as well. If we say, "I always get stopped by this traffic light," then most likely you will always get stopped by the traffic light. If you say, "I always get stuck in traffic," then you will always get stuck in traffic. If you fear that something bad is about to happen, then it becomes more likely that something bad will happen.

This would explain a lot about my family as I grew up. As I said in the first chapter, my grandmother was always certain that something bad was about to happen. If something bad wasn't happening right at that moment then it was just around the corner.

> **"Don't part with your illusions. When they are gone, you may still exist, but you have ceased to live."**
> **—Mark Twain**

It also makes sense to me from a scientific standpoint that visualization would be more effective and that perhaps emotions would be more effective in causing this phenomenon to occur. The amount of quantum bits that have to be set up to create the thought of the visualization and the emotions is larger than the amount of bits from just a simple fleeting thought.

This is easily understandable if you've ever sent an e-mail with a picture in it. An e-mail by itself is delivered quite easily and quickly. But an e-mail with a picture attached to it takes longer simply because there is more data involved. And it makes sense that if more quantum bits have to be arranged to generate the qwiff, then that qwiff is larger and more powerful in its interaction with the larger subsets of qwiffs within the Universal Qwiff itself.

It makes scientific sense then to implement the concept of the "vision board." Many life-coach types have suggested the benefits of the vision board for ages. A vision board is simply a poster that you make with pictures on it representing your goals. You sit down and meditate for a few tens of minutes a day while focusing your mind on the images of your vision board. For example, on my vision board I have made a movie poster saying

> **"One Day On Mars!**
> **Now showing in theaters everywhere!**
> **Based on the science-fiction novel**
> **by Travis S. Taylor."**

I focus on that image routinely. I focus on my wife and me all spiffed up—me in a tux and her in a slinky evening gown—getting out of a limousine onto the red carpet at the premiere of the movie. I also focus on how much fun we have at the premiere and the entertainment shows talking about what an awesome blockbuster it is.

I don't wish it would happen at all. And I don't think to myself that "it will" happen in time. I just fantasize through it. In my mind, and as far as the universe is concerned, I'm not waiting for it to happen, because time and space are all part of the universe. It did/has/will/is happening all at once. That is the same type of focus I had on becoming a scientist. I always just believed that I was/am a scientist and never let anything else cloud that fantasy, which became reality.

> "You can use your illusion let it take you
> where it may. We live and learn and then sometimes
> its best to walk away. Me I'm just here hangin' on,
> it's my only place to stay at least for now anyway.
> I've worked too hard for my illusions
> just to throw them all away."
> — Guns'n'Roses,
> "Locomotive" from *Use Your Illusions II*

My wife and I taught this to our four-year-old daughter. Every night before she would go to bed, she would focus on what she wanted — a swimming pool in our own backyard, of course. What else would a four-year-old want?

We told her not to "wish she had a pool" but instead to imagine "playing in a pool in our backyard." After months of visualization on her part, I received a much larger royalty check from one of my science-fiction novels than I had expected, and so my wife and I decided to put in a swimming pool in our own backyard. Needless to say, my daughter is a firm believer in *The Secret*. The universe provided for her visualization, and she swims in it every day.

> "Dream on, dream on.
> Dream yourself a dream come true!
> Dream on, dream on.
> Dream until your dream come true!
> Dream on, dream on, dream on . . ."
> — Aerosmith

Learning how to acquire the material things through *The Secret* is more along the lines of introductory skills. Everybody wants stuff. Think deeper about what true happiness is all about. My wife has a picture of a young healthy happy couple with two smiling children on her vision board. This is her desire of happiness—for us as a family to be generally happy and healthy. I cheat often and look at her vision board, too. I've trained my mind in such a way now that when I smile, an image, of my wife, daughter, son, two dogs, and cat pops into my head. When I think of that image a smile pops onto my face. The two things are connected now in my mind. My happy family and my happiness are the inverse transformations of my happiness and my happy family.

Your personal happiness I would say is an intermediate level of understanding the Law of Attraction. Perhaps there is a bigger scale to things where we learn as a collective mindset to think of good things. How easy is it for the world to believe we are having "bad times" when it is perpetuated around the clock on news shows? Are we letting these doomsayers on the television and in politics focus our minds on to negative things? What would happen if there was a positive message on the news every day and everybody on the planet focused on that for just a few minutes a day?

People in the general populace let their thoughts remain unfocused, wandering from qwiff to qwiff randomly. Hopefully, the lack of holistic focus is better than a truly negative focus. People watch train wrecks, but maybe they don't really focus on them. They are just receiving and not transmitting. This being said, the larger groups of positive thinkers that can be put together with common focus might help lead us toward a better end. One of my friends would say this sounds like a prayer circle.

But I don't mean we focus (or pray, if you prefer) for a person to get better or something so personalized. I'm talking about a focused group on scientific breakthroughs that cure diseases, disabilities, and old age. Louise Hay would likely say that you can heal yourself with the right mindset. She survived cancer and is now in her eighties and certainly doesn't look it. She says her focus allows her to live longer and stay healthy.

Could larger groups focusing their thoughts on the fountain of youth lead us to that end? Could larger groups focusing on a world of peace like that portrayed in *Star Trek*'s Federation lead us to world peace and in a joined humanity reaching out to the stars?

There have been "holistic experiments" in the past, but with questionable scientific methodology when determining if they were effective. There have been more recent studies performed by actual scientists that suggest such holistic approaches have an effect on things like crime rate and casualties in war. We'll discuss this more in the last chapter.

But what matters most, what you should take away from *The Science Behind the Secret*: Use your quantum brain wisely. Learn to focus on the way you think about everything. Remember, all your thoughts, even the fleeting ones, are quantum waves broadcast out through the universe. Don't waste them.

> **"Nothing has any power over me other than that which I give it through my conscious thoughts."**
> **—Anthony Robbins**

Chapter 10:

There Will Always be Skeptics So Don't Let them Get to You!

"When even the brightest mind in our world has been trained up from childhood in a superstition of any kind, it will never be possible for that mind, in its maturity, to examine sincerely, dispassionately, and conscientiously any evidence or any circumstance which shall seem to cast a doubt upon the validity of that superstition. I doubt if I could do it myself."
—Mark Twain

There will always be skeptics. My suggestion is to ignore them and pay them no mind. No matter what the idea, notion, theory, concept, or experiment, somebody is going to be skeptical of the outcome. Forget about those skeptics and move forward with your life. Let them be skeptical all they want. It is their nature; don't let it become yours.

Most of the skeptical comments, reviews, and debunking arguments against *The Secret* and the Law of Attraction use anecdotes of something bad happening to somebody in a bad situation and make the statement that, "Certainly they couldn't have wished themselves to be in this situation."

But *The Secret* is not a magic formula for transforming the entire universe. There are innumerable, complicated quantum wave functions affecting the Universal Qwiff, and our personal thoughts are only one very small ripple among the vast sea of quantum noise. But with practice and focus, we *can* have an effect. We can send out qwiffs to interact, entangle, and cohere with similar qwiffs. But we can hardly control the entire universe.

Another commonly cited skeptical argument involves the Olympics. Many of the elite athletes that participate in the Olympics use focus and visualization, i.e., *The Secret* by any other name. But in the end, only one athlete can win the race. If all of them were focusing with *The Secret*, why does only one athlete win the gold? That is the nature of any competition—only one person can win. And if the training and physical fitness are on equal footing, the winner will likely be the one who maintains their focus throughout training and throughout the competition. You often hear sports commentators talking about an athlete "just wanting it more," and that's why they win. This is hardly an argument against *The Secret*.

The idea that a quantum wave function is required for true consciousness—i.e., the quantum brain idea or Orchestrated Objective Reduction theory—has also met quite a few skeptics. In fact, many papers have been written discussing why a quantum brain isn't needed in order for consciousness to occur. Most of these skeptics have personal investments, whether they be money or blood, sweat, and tears, in the notion of artificial intelligence. Many of these skeptics simply want to believe that, with enough connections and enough software, human consciousness can be re-created, and that there's nothing too spectacular about how the brain and body must work in order for the consciousness of the mind to occur. But they've spent decades trying to develop artificial intelligence through the classical computational model, and have failed. Meanwhile, Penrose and Orch OR has shown great promise, and we are only at the very beginning of the testing and experimentation phase of this new science.

Thus far, all of the papers that I've read trying to discredit or shoot down the orchestrated objective reduction idea have been answered very handily.

The biggest skepticism of the idea is that inside the brain is too warm and too wet for quantum computation to take place. At the temperatures within the brain, it is argued, there is too much quantum noise for coherent quantum wave functions to be set up in the tubulin proteins for quantum computation to occur. But these skeptics were shot down by several papers showing that certain phenomena within these tubulin proteins and the microtubule-associated proteins could, in fact, make an ideal environment for quantum computation.

The same goes for *The Secret*. Forget the skeptics. And forget about those who would say to you that this is a scam and a waste of your time and money. We've shown there is plenty of potential basis within modern science and the understanding of the human brain and the universe for *The Secret* to take place.

> **"No Pessimist ever discovered the secrets
> of the stars, or sailed to an uncharted land,
> or opened a new heaven to the human spirit."**
> —Helen Keller

Some would say that it is human nature to be skeptical of things – of grandiose and wonderful ideas. The quantum mind idea and *The Secret* and the Law of Abundance and the Law of Attraction are great ideas. They're great philosophies from which to approach life. Even if we can't manage to prove they are 100 percent real, they are much better ideals to live by on a day-to-day basis than following adages like Murphy's Law or "'stuff' happens."

When I graduated from high school back in the '80s, it was standard practice for all students—following graduation ceremonies—to make their way as fast as they could to Panama City Beach in the Gulf of Mexico. Each year the graduating seniors have a theme at that beach vacation. And that theme always managed to make its way to T-shirts and billboards and bumper stickers, and into the hearts and minds of the young vacationing teens about to enter the real world. The year before I graduated high school (1985), seniors came back all excited about their beach trips; their theme had been "party naked." Sounds like a fun time.

The next year, when my friends and I graduated from high school and headed down to Panama City Beach, the theme somehow managed to be "stuff happens." Well, it wasn't "stuff," but you get the idea.

And maybe it was from my own perception of that trip, but I observed that everybody at Panama City Beach that summer seemed overly belligerent and bordering on violent as opposed to happy and excited.

I blame it on the theme itself. That theme changed people's frame of mind, and that theme became an excuse for their behavior. I recall seeing one quite intoxicated fellow punch another guy in the head completely out of the blue, and then say, "'Stuff' happens, man!"

I'd much rather have had the theme from the previous year.

I know this is just another anecdote, and a lighthearted one at that, but it's a personal observation and experience that seems appropriate, nonetheless. A simple frame of mind based on a few simple words changed the attitude of the students on vacation. Those two words were so powerful that it shaped the emotions and actions of an entire crowd of beach-hopping graduating seniors.

Words are powerful—very powerful. One of the things we teach our daughter to say to herself often is that "everything we do and everything we say is important." With "stuff happens" and "party naked" in mind, we should be mindful of our words and choose them carefully. A global skepticism is unhealthy. Think of the emotions usually triggered by the dark skeptic's words. Even more harrowing, look at the economic troubles of 2008 and 2009, where lack of consumer confidence was repeatedly cited as a leading factor in causing the sharp economic decline; the economy failed because nobody believed in it.

> "Words have the power to both destroy and heal.
> When words are both true and kind,
> they can change our world."
> —Buddha

Think of how having a frame of mind following the teachings of the ideals and philosophies of *The Secret* might simply change your attitude on a day-to-day basis. Just the positivity of attitude alone triggers emotions as different as "party naked" versus "stuff happens." If more people had a similar frame of mind, think of the holistic improvement in attitude that could occur.

So skeptics be damned, I say. *The Secret* is a good axiom to live by and is a great frame of mind to have.

> "I am not very skeptical . . . a good deal of
> skepticism in a scientific man is advisable to avoid much
> loss of time, but I have met not a few men, who . . .
> have often thus been deterred from experiments or
> observations which would have proven servicable."
> —Charles Darwin

I don't believe it's human nature to be skeptical—curious and inquisitive, certainly. And there is diversity in human nature, so skepticism may come naturally for some people. But if it were all of humanity's nature to be skeptical, then mankind might have never ventured out of Africa, across continents and across oceans and across the vastness of space to the Moon, and in the future to Mars and other planets, and maybe even to the stars.

People were skeptical of the Vikings and Christopher Columbus and Lewis and Clark and Jim Bridger and Livingstone and Stanley and Alan Shepard and Neal Armstrong and Buzz Aldrin and Michael Collins. People were skeptical of Einstein and Heisenberg. One can just go read the few negative reviews of *The Secret* and of Louise Hay's books at amazon.com to see the skeptical viewpoints.

There is a difference between healthy scientific skepticism and simple negative skepticism. Scientific skepticism is a must in maintaining the validity of new scientific theories and experiments. We must be professionally skeptical of the new ideas in such a way that we test them, study them, and re-test and re-study them until we are certain that they are truly valid scientific ideas. This type of skepticism is healthy for humanity.

> ""It is not the critic who counts; not the man who points out how the strong man stumbles, or where the doer of deeds could have done them better. The credit belongs to the man who is actually in the arena, whose face is marred by dust and sweat and blood; who strives valiantly; who errs, who comes short again and again, because there is no effort without error and shortcoming; but who does actually strive to do the deeds; who knows great enthusiasms, the great devotions; who spends himself in a worthy cause; who at the best knows in the end the triumph of high achievement, and who at the worst, if he fails, at least fails while daring greatly, so that his place shall never be with those cold and timid souls who neither know victory nor defeat."
> — Theodore Roosevelt
> (April 23, 1910)

Skepticism, whereby people simply make their living as professional skeptics or are negative from a general standpoint on most things, is not healthy. The "doomsayers" who say bad things evoke emotions, feelings, thoughts, and actions that are often detrimental to society. They could learn a lesson from Thumper's mom in *Bambi* and keep their mouths shut. You should take the same lesson and ignore the general negative opinions, messages, and so on.

One night while writing this book, I watched my local news because I wanted to see the weather. For the first fifteen minutes of the program, all they said was how bad the economy was, how many people were robbed, mugged, raped, or killed, followed by a rundown on local criminals and how bad our traffic was getting. And, of course, our traffic, according to the news, was only going to get worse.

I live in a fairly small town in north Alabama. People here are generally good-natured Southerners, good neighbors. There had to have been something good, cool, unique, great, or even awesome that happened in the town, but there was no mention of it. I used to think that it was educational to watch the news, but from recent study with more open eyes I am beginning to think that the evening news is not good for anybody. I'm turning off the tube and looking up my weather information on the Internet for now on. I would class the programming directors of these news programs as doomsayers. Perhaps they don't realize what they are doing. Perhaps they do. Certainly they realize that we have become a nation of people within which train wrecks sell.

> **"A person with a new idea is a crank until the idea succeeds."**
> **—Mark Twain**

A broader, more sinister face of the skeptics, negativists, and doomsayers is represented by politicians who do their level best to convince one group or another (a voting bloc, usually) that the world is coming to an end if we don't pass some law or ordinance or take some action. They tell us that the economy is bad and getting worse, that we are running out of stuff (oil, energy, medicines, and sometimes even water—on a planet that is mostly water!), and that the world is doomed if we don't continue to elect or promote them. This is a deeper level of holistic skepticism that isn't helping us.

Forget them. Only let the positive aspect of life, the happiness of being, pervade your mind. At the very least, watch the news and politicians with guarded reserve.

The great adventurers, philosophers, and thinkers did not let the skeptics dissuade them. I believe it's human nature to be *great*. It's all of the nonsense that we let other people pour into our minds that sometimes keeps us from being that way. We have to unlearn the skepticism. Ignore it and push away and absolutely resist those that would be negative on a routine basis. There is no place for it within humanity.

> **"New and stirring things are belittled because if they are not belittled, the humiliating question arises, 'Why then are you not taking part in them?'"**
> **—H. G. Wells**

So, let us address one of the skeptic's debunking approaches to *The Secret*. My favorite approach used by the skeptics is one involving soldiers in combat.

The sample argument typically claims that if a soldier is killed in combat, it wasn't *his* thoughts that caused his reality. He was ordered there by other people, and then shot at by people from the other side of the conflict. Any amount of wishful thinking certainly wouldn't have saved his butt from the meat grinder of combat.

But this argument is in complete agreement with the idea of *The Secret* as far as I can tell. Here is why. What is the mindset of the good soldier? They are trained to do a job, a very hard job, a very dangerous job, a very heroic job. That soldier may have joined the military for whatever reason, but the long line of events (and thoughts and motivations) throughout that person's life led them down the path of joining the military, and eventually led to a military situation: combat.

When soldiers go into to these combat situations, they've already had the thoughts that led them to where they ended up, and in fact are exactly where they expected to be. Not that it would be too late to implement change through *The Secret* while deeply involved in the combat situation, but it would prove quite difficult. At that point, you would be in the competition environment, and there is the aspect that the best physical and mental athlete is going to win. The best trained from all aspects of life will be the most successful. There is no argument in this approach that debunks *The Secret*.

Whatever the debunking argument might be, crime victims, car accidents, being dumped by their girlfriend, or pure happenstance of just being in the wrong place at the wrong time, the arguments are just anecdotes. They are not logical proofs, scientific evidence, or mathematical basis for bunking or debunking. These arguments are used to evoke the emotion of doubt without any real evidence one way or the other. Here is where we need to apply the healthy skepticism of scientific methodology to debunk the darker and more sinister skepticisms.

Again, *The Secret* isn't about maintaining absolute control over the entire universe. It *is* a way to interact, entangle and cohere positive thoughts and events via the larger Universal Quantum Connection—i.e., your positive thought qwiffs cohering with similar qwiffs in the universe at large.

But instead of focusing on the darkness, let's focus our efforts on the brighter, happier side of things and simply test our theories. The outcome will shine light on the darker corners of humanity and wash them out.

> "It is really quite amazing by what margins competent but conservative scientists and engineers can miss the mark, when they start with the preconceived idea that what they are investigating is impossible. When this happens, the most well-informed men become blinded by their prejudices and are unable to see what lies directly ahead of them."
> —Arthur C. Clarke

Chapter 11:

The Proof is in the Pudding!

> "If we knew what it was we were doing,
> it would not be called research, would it?"
> —Albert Einstein

> "When the province of physical theory was
> extended to encompass microscopic phenomena
> through the creation of quantum mechanics,
> the concept of consciousness came to the fore again.
> It was not possible to formulate the laws of quantum
> mechanics in a fully consistent way
> without reference to the consciousness."
> —Eugene Wigner

Are there scientifically valid experiments that we could conduct to prove that the Law of Attraction is real? Perhaps. The status of modern quantum theory, Orchestrated Objective Reduction, and neurology, is at an early stage. We do know for certain that the quantum-physics phenomenon is real. We don't know how strong it is and what level of implementation we can attain with it, but it exists.

Possibly, there is no limit to what can be done with it, but we need to make some scientifically valid assessments before we come to any conclusion that can be accepted by mainstream science.

With that in mind, there have been several studies and experiments in many different topics of study that support *The Secret* as a real phenomenon. Let's discuss some of those.

At the heart of *The Secret* is the quantum-physics theory of Orch OR and the quantum brain. Does the brain really function like a quantum computer?

After decades of failure in modeling the brain on classical computers, Orch OR and the quantum brain are the first truly encompassing theories that might describe why and how consciousness occurs, what thought really might be, and how thought interacts with the universe.

At the basis of Orch OR is Roger Penrose's addition to quantum mechanics, which defines the level of energy where an object can no longer be in multiple quantum states at the same time. This level is the quantum gravitation limitation. We discussed how any object could possibly be in multiple quantum states at the same time, but the length of time this could happen is inversely proportional to that object's "gravitational self energy" or mass.

Since photons (the quantum thingies that make up light) have no mass, based on Penrose's theory one could expect that a single photon could stay in multiple quantum states an infinite amount of time. This, of course, is just a theory.

On the other hand, we see the multiple quantum states of single photons occurring in Young's double-slit experiment all the time. Young's double-slit experiment has been conducted with larger particles and has shown that these particles can also be in multiple quantum states at the same time. So it seems viable.

But nobody has conducted the Schrödinger's cat

experiment with particles the size of a cat. The largest scale of the Schrödinger's cat type experiment was when the physicist Anton Zeilinger of the University of Vienna fired a single buckyball, which was made of sixty carbon atoms, through two slits at one time. This demonstrated that an object at least as big as a buckyball could be in two quantum states at once. It went through both slits at the same time. The buckyball was big, but still, no Schrödinger's cat.

Penrose has the idea of using a very tiny mirror—one that is a little smaller than a human hair but is still a macroscopic-scale object—and experiment with it in such a way that its position is uncertain and represented by two different quantum states. If Penrose's quantum-gravity equation is correct, then very quickly the mirror will settle into a single quantum state. If this experiment is conducted and achieves the results expected, then the nearly century-old dilemma of Schrödinger's cat will be solved. Such an experiment will lend much credence to Orch OR and the quantum brain concept. And most important of all, we can finally let that poor cat out of that box!

Another experiment has been proposed that could also support Penrose's theory. The physicist Andreas Mershin is planning to test the idea that quantum effects take place in the tubulin proteins in the brain. Penrose suggested that tubulin could be co-located on the surface of a plasmon: a piece of metal with a group of electrons acting in a certain way together. The hopes are that the tubulins and plasmons will become entangled with each other in such a way that the electric constants of the tubulin will be altered. If the tubulin properties are altered in this way, it will mean that tubulins are capable of becoming quantum entangled with other particles. And that would give a lot of credibility to the quantum-brain theory. Few, if any, other consciousness theories have such properties that can be tested in laboratory experiments. This is part of the beauty of Orch OR. It is testable.

> **"In order to put meaning back into our lives,
> we should recognize illusions for what they are,
> and we should reach out and
> touch the fabric of reality."**
> —Evan Walker,
> *The Physics of Consciousness:*
> *The Quantum Mind*
> *and the Meaning of Life*

Recent observations by the physicist Jack Tuszynski of the University of Alberta have also suggested that the brain tubulins work at a quantum level. He is studying a famous anticancer drug known as Taxol. Taxol binds to microtubules that are responsible for cell division and prevents them from dividing. In cancer research, the goal is to prevent only the cancer cells from dividing and spreading. Through observations it has been shown that Taxol binds to the microtubules with a small amino-acid arm. Tuszynski describes these arms as docking mechanisms.

As biochemists alter the drug Taxol with hopes of making it more effective, they have observed that the altered drugs require two of these docking arms in order to connect to the microtubules. These two docking arms are only a fourth of one billionth of a meter in diameter. Anything that small is about the diameter of an atom and is therefore clearly operating at the quantum level. Something within the microtubules is functioning via quantum interactions. That certainly seems to indicate that Orch OR is real, and that our brains could truly be quantum computers.

In the field of consciousness study, we're likely to see even more exciting experiments, observations, and discoveries in the very near future. As Orchestrated Objective Reduction becomes more accepted by mainstream scientists, experiments aimed at consciousness theories will become more accepted as well.

> "See I could not afford to allow anything to come into my mind that would distract me from my goal and from my vision. Well I'd set the goal to walk out of the hospital on Christmas. That was my goal. Eight months later I walked out of the hospital on my own two feet. They said it couldn't be done. That's a day I will never forget."
> — Morris "The Miracle Man" Goodman
> from *The Secret*

There is other evidence supporting the quantum-brain theory and the notion that *The Secret* and the Law of Attraction are working through our quantum brains. And that is in the area of healing through visualization.

As far back as 1988—in a graduate thesis written by Lisa Neumann Earle from Florida Atlantic University—the effects of visualization on the healing process were investigated. In that experiment, data was obtained from forty-two different subjects between the ages of thirty-one to seventy-nine-years-old, all of whom were suffering from life-threatening diseases, such as cancer, heart disease, and others.

The patients were trained in the use of mental imagery and were put through a series of interviews to gauge their pain and comfort levels both before and after the visualization experiment. The patients and their doctors were interviewed before, during, and after the process.

The results of the study showed that diet, exercise, medication and vitamins, and the standard treatments alone created no unexpected statistical variations. However, the patients undergoing the mental imagery showed a major improvement in both attitude and condition. The major final recommendations of the thesis were that larger studies needed to be conducted, mental imagery might be useful in patients with other diseases such as AIDS, and mental imagery might be a good tool for maintaining good health and quality of life as well as useful with healing.

In 2006, a paper appeared in the *Journal of Sports Sciences* from scientists affiliated with University of Western Ontario and the School of Sport, Health and Exercise Sciences, University of Wales, titled "Imagery Used by Injured Athletes: A Qualitative Analysis."

In this paper, the scientists took ten injured athletes who were already receiving physiotherapy at the start of the study. The athletes were interviewed about their injuries and the rehabilitation process. The athletes were taught to visualize and create mental images of their injury healing itself and of the pain dispersing away from their bodies. An example of such an injury rehabilitation visualization would be to imagine a broken bone getting back together and melding back into a solid healthy bone. Similar images were used to visualize healing muscle tears. The results of this study were:

"the implementation of imagery alongside physical rehabilitation should enhance the rehabilitation experience and, therefore, facilitate the recovery rates of injured athletes."

Another paper from the *Journal of Imagery Research and Sport and Physical Activity* titled "Imagery Used During Rehabilitation From Injury," showed that athletes who implemented mental visualization during the healing process healed significantly faster than athletes who did not. Both of these studies revealed that visualization of the healing process speeds up and facilitates healing.

This isn't science-fiction; these are actual scientific studies that have been talked about in most of the popular fitness magazines such as *Runner's World* and *Men's Fitness*. And it is very similar to how Louise Hay describes self-healing in her teachings.

> **"I am strong and sound. I am well structured."**
> — Louise Hay

And more than just for healing, mental imagery is also being studied for performance enhancement. In another recent study published in 2006 in the *Journal of Sport Sciences* it was shown that visualization of muscle stretching while actually stretching increased muscle flexibility significantly. During this study there were three groups: one only visualized stretching; the second visualized stretching while they stretched; and the third only stretched and did not implement mental imagery. Both of the first two groups showed significant increases in flexibility that were higher than the third group. The groups that used mental imagery also reported higher levels of comfort during actual stretching.

Another study in the *Journal of Physical Therapy* in 2005, titled "Can Mental Practice Increase Ankle Dorsiflexor Torque?" measured the effects of mental imagery on the increase of muscle force. Twenty-four subjects were randomly chosen and assigned to either a physical practice group, or a mental practice group, or a control group. The practice groups either actually did a physical strength-training exercise or (if they were in the mental group) mentally practiced and only visualized doing the exercise. The study showed that both practice groups saw significant increases in muscular force while a control group that did nothing did not. Simply visualizing the exercise increased the strength of the mental practice group in measurable amounts. Similar studies have been conducted to test the improvement of balance in the same way through visualization and have observed the same types of results.

How is it that simple mental imagery has such an impact on how our bodies function? We know these microtubules are part of our cells and are involved in cellular division. We've also discussed that the microtubules might be the heart of the quantum activity within our brains and where our thoughts take place. We've also shown how quantum computation is important through how our eyes work, which suggests quantum phenomena at work when we have visual experiences. We also know that quantum things that become entangled are connected over any distance and interact with each other instantaneously—Einstein's "spooky action at a distance."

All of the cells in our bodies came from that very first miracle group of cells's at conception. So every cell in our bodies should be quantum connected to every other cell in our bodies. A visual thought requires setting up many quantum bits within our brains that then propagate through the universe as a quantum wave function, or qwiff, as we have discussed. This qwiff will likely cohere with other qwiffs that are most like it. And it only stands to reason that the qwiffs representing the cells' quantum states within our own bodies would be quite similar. It makes sense that by visualizing cells in your own body performing in a particular way, you will help cohere the reality of those cells performing as visualized.

The ramifications of such a concept are mind-boggling. Simple visualization has been shown to increase strength, speed up healing, and increase attitude and comfort levels.

And we're just now at the tip of an iceberg. As we learn more about how to visualize the way we want our own bodies to feel and perform, we may realize that we truly can heal our own lives and enhance them simply through the right process of thought.

Don't take this the wrong way, or take it too far, and lead yourself to the conclusion that we don't need doctors or medicines or surgeries or dentists and so on. We're likely to create our own limits within our minds and will need such professional healers for big things until we learn how to have no limits to what we believe we can do within our own minds.

> "The positive effect of kindness on the immune system and on the increased production of serotonin in the brain has been proven in research studies. Serotonin is a naturally occurring substance in the body that makes us feel more comfortable, peaceful, and even blissful. In fact, the role of most anti-depressants is to stimulate the production of serotonin chemically, helping to ease depression. Research has shown that a simple act of kindness directed toward another improves the functioning of the immune system and stimulates the production of serotonin in both the recipient of the kindness and the person extending the kindness. Even more amazing is that persons observing the act of kindness have similar beneficial results. Imagine this! Kindness extended, received, or observed beneficially impacts the physical health and feelings of everyone involved!"
> — Dr. Wayne Dyer

> "There are no accidents."
> — Master Oogway from *Kung Fu Panda*

I experienced a similar use of mental imagery during a sports-injury rehabilitation of my own. While training for a marathon, I began experiencing extreme heel pain one day after a twenty-mile run. The pain continued to get worse, and after studying the problem and seeing several physicians, it was clear that I had developed plantar fasciitis. This is a very painful condition in which there are tears or at least inflammation that occurs in the tendon that stretches across the bottom of the arch of your foot. The standard treatment for plantar fasciitis is to stay off your feet, wear special orthotic arch supports, foot massages, ice, anti-inflammatories, and use a special splint that keeps your foot flexed at night when you sleep. None of these things worked at all for me.

I was reaching the point where I was beginning to consider the more invasive treatments explained by several different sports-medicine doctors of cortisone injections and surgeries. The other alternative was actually not doing anything for six months or more. None of these ideas seemed like good ones to me.

This is when I discovered the sports-injury visualization studies. In every idle moment, I began visualizing the swollen part of my arch shrinking like a balloon deflating and the tendon looking healthy and happy and me running and feeling great. I also visualized the tendon weaving itself back together in the spot where the MRI suggested that it was damaged. I focused on this very regularly for about two weeks and began to feel some improvement. Maybe it was all in my head, but I didn't really give a damn. It felt better one way or the other.

One morning after visualization, I decided to surf the Internet for a marathon later in the year to allow more time for healing. And as the browser opened up, a pop-up window opened in the middle of the screen claiming to have a cure for plantar fasciitis called "heel-that-pain."

Just for fun, I opened up the Web site and read what they had to say and decided what the heck. The product being advertised was less than $30, and I'd already tried no telling how many thousands of medical insurance dollars' worth of cures that didn't work. What would it hurt to spend another thirty dollars plus a few for shipping? I ordered the little inserts being advertised.

A week later they came in the mail, and I immediately stuck them in my shoes. In less than two hours, the pain in my arch significantly decreased. I couldn't believe it. In less than a week, my plantar fasciitis was barely noticeable. In three weeks, it was completely gone! I've run several marathons since with no pain from plantar fasciitis.

Now this isn't an advertisement for that particular product.

Instead, this anecdote inspires a question. It seems likely to me that my visualization of the healing process aided my actual healing. To me, the real question is: Did my visualization of my plantar fasciitis being healed propagate as a qwiff throughout the universe that in turn cohered with other qwiffs that led to me finding these little inserts?

Of course, the skeptic would say that marathon Web sites are a great place to advertise for devices to help with running-related injuries. But nevertheless, a solution did present itself. A solution did the work following my visualization and not before. It at least makes you stop and think.

> "Bring it! You can do anything for twenty seconds!"
> —Tony Horton *P90X* Fitness Trainer

As a side note, for each marathon I run, from the start of my twenty-week training plan to the finish line, I visualize the time clock as I step across the line (I did this back when I was mountain-bike racing also). If I hurt during training, I recall the powerful words of great trainers, coaches, and sports heroes throughout history and say those words of wisdom to myself.

I also chant in my head kick-butt phrases in my best Dick Vitale or Macho Man or Ray Lewis inner voice. It seems to help. If that doesn't work, I crank the volume of the iPod up a bit and play some headbanging rock 'n' roll and occasionally Lance Armstrong's voice breaks in urging me to keep pushing. The power of words is great and can help us achieve things we never thought we could.

> "Empty your mind, be formless, shapeless—
> like water. Now you put water into a cup,
> it becomes the cup, you put water into a bottle,
> it becomes the bottle, you put it in a teapot,
> it becomes the teapot. Now water can flow or
> it can crash. Be water, my friend."
> —Bruce Lee

On a much more grandiose scale of implementing the teachings of *The Secret*, physicist Dr. John Hagelin and others have actually been setting up large groups of people trained to visualize and focus their minds on specific topics. Following the visualization efforts, scientists then study societal statistical data for any impact.

The idea of these holistic experiments is to have the mental-focus groups visualize peace, after which the scientists review the statistical data of crime rates, accident rates, war and terrorism, and other indicators for any reduction. According to Haglin's Web site, permanentpeace.org, there've been over fifty such experiments, with the data published in over twenty-three different scientific studies.

If the studies are correct, as Hagelin believes, the status of the world is due to the holistic mindset of humanity. Cultures really might be getting what they asked for. A lot of this concept stems from the Vedic tradition, and Dr. Hagelin believes quantum physics and his experiments have validated the following four ideas:

- There is a single unified field of intelligence at the basis of nature
- Humans can experience this field
- Learning to experience this unified field of intelligence will generate rapid individual growth
- A holistic radiating influence of peace over humanity will occur.

The first bullet about this single unified field of intelligence being the basis of nature is very similar to Penrose's concept that there must be some underlying non-computable quantum laws as the building blocks of our universe. This also ties in to the idea of Wheeler's universal quantum consciousness.

Again imagine that there is one very complicated universal qwiff that includes everything. Each individual consciousness is simply a subset of the larger universal qwiff. It stands to reason that if subsets of the universe are conscious, then on a much larger scale, so is the universe. When our consciousness leaves our bodies at death, what happens to the subset qwiff explaining us at that point? It's still a part of the universe. Where does our consciousness go?

The second bullet suggesting that humans can experience this field is intriguing. We've discussed how our brains really are more than just a quantum computer, that they are truly a transceiver of qwiffs: we can send and receive qwiffs to and from the universe. The problem is, how do we know if we are receiving one?

I proposed earlier that receiving outside qwiffs might be the reason for random thoughts and weird dreams. It is possible that other weird phenomena between twins, déjà vu, and other ideas outside mainstream science might have a mechanism through which they could function.

Is it possible that if we truly learn to focus our minds and quiet our minds, we could cohere with the universe and actually—to put it in the parlance of the Jedi Knights of *Star Wars*—hear the Force speaking to us?

**"In the beginning of practice, we focus primarily
on learning the form of meditation.
We learn how to work with our bodies so that
we can sit comfortably in zazen with good posture
and natural breathing, and we learn how to work with
our minds so that we can focus our attention.
At some point we experience an opening to our true nature.
We call that first experience of nonduality
'the shift' because it is a shift from our usual way of
perceiving reality. Suddenly, instead of perceiving
only the apparent reality, what appears to be so,
we have a direct experience of the absolute reality
that underlies and pervades all things, including the self.
Once we have glimpsed the absolute,
the self never again seems so solid or permanent.**

> **Our usual way of perceiving the self and the world has been cut through, even if only for an instant, and it becomes easier to remember that duality is just one side of reality."**
> —Genpo Roshi, *The Path of Being Human*

There have been people such as Jane Roberts who claim to have "channeled" entities from somewhere "out there," entities claiming to be energy beings. The being Roberts claims to have channeled is called Seth, and it apparently taught her all about the nature of the universe and reality. Esther Hicks is another person who claims to have tapped into energy entities that give guidance and describe reality. She claims to have learned, after many days of meditation, how to quiet her mind and communicate with the collective entity known as Abraham. Her myriad of books on the Law of Attraction are based on her sessions with this Abraham.

Roberts and Hicks tell stories that are not unlike the story of Buddha, who meditated under the fig tree for so many days. I don't want to debate here whether Jane Roberts, Esther Hicks, or Buddha actually reached some new level of mental focus and enlightenment or not.

But I do believe we have firmly established a possible mechanism for such a thing through quantum physics.

Again, I'm not endorsing or debunking the idea. I am merely putting forth the suggestion that if there is an underlying quantum consciousness that is universal, and if human brains are quantum transceivers, then it is possible that human-to-universe and universe-to-human quantum transmissions occur. In fact, it's entirely plausible.

The third bullet is sort of self-evident. If the first two bullets are true and we learn how to truly understand them, then clearly the means for rapid personal growth would be at hand. There's no reason to belabor the point.

I would like to ask you to think about one thing, though. Where do new ideas come from? When is it that an idea is truly original or it is merely the next logical step of the previous concept?

Anybody who has ever had a eureka moment realizes that somehow a new idea nobody has ever had before manifested itself within their mind. How? Why? Where did they come from? If there is a basic universal reality or intelligence that makes up the universe, is it possible that the eureka moments are when you set up just the right intricate quantum wave function in your brain? Is it similar enough to a part of the universal truth that it cohered as the qwiff received by your brain and experienced as your eureka thought? Eureka moments might really be glimpses at the universal truths.

For most scientists and thinkers and experimenters, their eureka moments took years of thinking, studying, and experimenting with the universe to achieve. Sometimes, they might never come. A reason for this might be that the universal truth that the scientist is trying to glimpse is so complex that it is difficult to "wrap your mind around the concept." To broadcast just the right qwiff to the universe probably takes years of honing the idea so that it will correlate and cohere with that piece of universal qwiff.

Of course, those folks who could implement the second bullet and just communicate with the universe would be able to attain such an enlightenment through meditation. Seems a lot easier than all those years of graduate school!

The final bullet is the most exciting. If the first three bullets are real, and humanity could truly shape its reality from a holistic perspective, then with enough people focusing their quantum brains on the same visualization, large-scale interaction with the universal qwiff should be observable. If we want war, we're likely to get more of it. If we want peace, that is what we should holistically focus on. But, as stated before, we shouldn't have antiwar rallies because the focus there is a negative, i.e., "war." Instead, we should focus on peace and happiness. (And that perfect beer, and the one day work year, and spaceships that can take us to the stars.)

※ ※ ※

Perhaps there are simpler experiments that we could envision that would lead us to a more solid statistical and verifiable proof that *The Secret* is real. We've shown the mechanisms needed for it to be real and have listed a lot of evidence in support. We just have to go that next step and prove it beyond any doubt to everybody within humanity.

Perhaps we could use the large mental-focus groups to create a single image in their minds like everyone in the world seeing something "purple." Who knows how that might manifest itself in reality? If all of a sudden the next Windows operating system was called Purple OS or the next iPhone was called the Purple Phone or the hottest show on television was called Purple (you get the idea), then we might be on to something. That type of experiment might be more simple to measure and understand than other broad-spectrum experiments.

This is sort of the same idea that John Hagelin discusses in *The Secret* and on his Web site. It's just that the data they are analyzing is very complicated data, and it is harder to convince people who are skeptical that there are real trends generated by the mental focus teams. But that's okay. Science is hard, and the proof is in the pudding!

We've identified here that there might be a bowl of pudding out there somewhere. We need to find it. Eat it. Absorb it. Understand it. And then shout from the rooftops that we all are part of something bigger. Something waaaaay bigger. Through the science described in this book, humanity's future is becoming clearer every day. Our future is that we must realize and understand that we are all a part of *The Secret*, the Law of Attraction, and the Universal Quantum Connection!

Chapter 12:

Final Thoughts

**"That's one small step for a man.
One giant leap for mankind."
—Neil Armstrong,
the first human to walk on the Moon**

I've taken you on a journey following my path to the level of awareness that I've reached thus far. And that level of awareness is one of accepting that the universe is bigger than humanity tends to accept. And not only is the universe bigger, but it is smaller at the same time. We're all connected: through ideologies, religions, philosophies, modern physics, and—at the core of it all—the science of the brain.

As far back as three thousand years before the Current Era, there were great thinkers teaching the ancient societies how to achieve their goals, desires, inner peace, and enlightenment. Throughout human history, many of these great thinkers taught very similar ideas about how to achieve these goals. And we've shown how these great ideas evolved into what is now, in the modern world, called *The Secret* and the Law of Attraction.

Understanding *The Secret* and how to implement it for enriching one's daily life is a good thing. It's great to realize that you can implement *The Secret* to live the life you want.

However, the real enlightenment comes through looking deeper into how and why it works. Why?

Do you really need to know how your iPod works in order to enjoy the benefits of having one and using it? Do you really need to know how a hammer is constructed from raw metal and wood, how the handle is milled from lumber and the head is blacksmithed from ore and then assembled, in order to enjoy using a hammer? Of course not.

But those are simple tools.

The Secret is different, because it is more than just a tool. It lies at the very heart of humanity and how we are connected to the larger whole of the universe.

Understanding that connection with the larger whole is a level of enlightenment that allows us to feel better, both about ourselves and about how we fit into the grand scheme of things. Everything in the universe is all part of the universe.

Remember we're not just in the universe. The universe is a large mishmash of all of the energy vibrations, the quantum wave functions, of everything within it. Every individual person, every squirrel, every tree, every rock, every hydrogen atom is a small ripple. And when those ripples are added together, they make up a ripple that is much greater than the sum of its parts. We are the universe.

I hope you found useful tools to add to your own personal philosophy and way of thinking. But more important, this isn't just another book on how to use the Law of Attraction. There are plenty of those. If you found a deeper understanding of what the Law of Attraction is, of what *The Secret* is, and how it fits within our reality, rather than simply seeing this as a manual with tools to gain wealth and other desires, then we've succeeded.

Don't get me wrong, I think gaining wealth and all of the things you desire is a wonderful goal, so wonderful that I think I'll DO it myself. However, the philosophy, the science, and the sheer magnitude of the Universal Quantum Connection is a piece of knowledge that is just as enriching in itself.

The late science-fiction writer Douglas Adams had a whole series of books called *The Hitchhiker's Guide To The Galaxy*. In those novels, he described how various species would become sentient and crawl up out of the marshes, and that they would all tend to ask the same question.

"What is life?"

And the response from the older and wiser civilizations would be something along the lines of

" . . . life is big and round and yellow, sort of like a grapefruit."

Of course, Adams was a humorist first, but that description is more than just a silly statement. Adams seems to suggest that life just *is*; there is no deeper meaning. That is a lonely view of life. This would suggest that life is merely a random occurrence and disconnected from the rest of the universe and everything in it.

Of course, I strenuously disagree.

The philosophies and ideas and science we've discussed throughout this book refute this. Life *is* much bigger, much rounder, probably not yellow at all, and certainly not much like a grapefruit.

Life is connected to all other things—all other life, not to mention every other object in the universe, from the beginning of time through the present to the end of the universe—through quantum physics. The underlying noncomputational truths of the universe could be the unknown driving force of life, the universe, and everything.

Through *The Secret*, the Law of Attraction, and the Universal Quantum Connection, life, is connected, driven, and should be abundant.

"There is a difference between knowing the path and walking the path."

— Morpheus
from The Matrix

Afterword

Can conscious thoughts directly affect the world around us? Are our minds in tune with other beings and the universe? Is there a deeper reality of cosmic wisdom?

Ancient philosophies, religions, spiritual traditions, and popular culture, including the film and book *The Secret* assert such effects can and do occur. But for most scientists the answer to these questions is a resounding and emphatic NO! Conventional science sees conscious thoughts and feelings as computations among brain neurons, no different than interactions among silicon bits in modern information technology. In this view, conscious thought could no more influence distant events than could a laptop or iPhone directly change the Moon's orbit around the Earth. Could science be wrong?

As Travis Taylor tells us in this book, the answer is YES. Within limits, it is scientifically plausible that conscious thought can shape reality through quantum effects in the brain.

Quantum effects? Our everyday world is governed by "classical" physics such as Newton's laws of motion and Maxwell's electromagnetic equations, predicting almost perfectly the measurable behavior of matter and energy. In classical physics, particles or physical objects are viewed as separate and distinct entities interacting predictably, like billiard balls. Based on classical physics, scientists in the late nineteenth century believed everything about physical reality was explained. But a few seemingly minor glitches unraveled their tidy view, uncovering the weirdness of relativity and quantum physics.

"Quantum" means, literally, the smallest unit of anything, like fundamental units of energy, or matter, generally at extremely tiny scales. But behavior at these scales is so bizarre that the legendary physicist Richard Feynman famously said, "Anyone claiming to understand quantum physics is either lying or crazy!"

For example, quantum particles can be in two or more states or locations at the same time! An atom can be both here and there simultaneously. At the same instant an electron can have opposite spins, a photon opposite polarizations, and an information bit both 1 AND 0 states (a quantum bit, or qubit). Quantum particles exist not as billiard balls, but as wavelike probabilities described by a quantum wave function, or "qwff" (pronounced *qwiff*).

Even more strange, schizoid qwffs separated by distance and time remain intimately connected by "non-local entanglement," what Einstein termed "spooky action at a distance." Perturb a quantum particle, and its entangled twin responds instantaneously, wherever it is. Spooky, yes, but entanglement has been verified many times over, reaching across oceans and bouncing off satellites.

In quantum computers, qubits interact by entanglement to compute, with potentially dazzling efficacy. Entanglement is a deeply mysterious connection among spatially separated quantum states and particles. It has no place in classical physics.

Quantum computations synchronized to brain electrical rhythms have been linked to consciousness and—via entanglement—suggest the potential to shape one's reality. Entanglement between minds of living beings could form the ... of love and emotional connection, not to mention the ... action as described in *The Secret*.

... quantum property is coherence, or condensation, ... articles become unified and act collectively as a single larger entity. Consciousness, and life itself, may utilize quantum coherence.

But we generally don't perceive quantum strangeness in our everyday experience. Reality seems divided into two worlds, defined respectively by quantum and classical laws. The separation is not just in size. Though rooted in subatomic physics, some quantum effects, including entanglement, occur at large scales. Indeed, the boundary, or edge, between quantum and classical physics remains a major mystery, one in which consciousness appears to play a key role.

Experiments by Niels Bohr and other physicists at the turn of the twentieth century seemed to show that quantum particles lose their qwff-like state of multiple possibilities only when consciously observed. When looked at, multiple possibilities instantaneously reduce, or collapse, to one definite state. Bohr and others concluded consciousness "collapsed the wave function," with the particular choice of classical state seemingly random. This view became known as the "Copenhagen interpretation," after Bohr's Danish origin, and implied that consciousness is, to some extent at least, capable of choosing or influencing perceived reality.

But there are serious problems with Copenhagen. For one, it placed consciousness outside science. And what about *unobserved* quantum possibilities? How large might they become, for example, if amplified by a quantum system? Is everything smeared out into possibility waves when hidden from conscious observation?

In 1935, Erwin Schrödinger posed his famous thought experiment, still known today as Schrödinger's cat. If the fate of a cat in an enclosed box is tied to a qwff-like quantum particle in multiple possible states, according to Copenhagen the unfortunate cat would be both dead and alive until someone opened the box and had a look. Schrödinger intended to ridicule Copenhagen, to show how absurd was the notion that consciousness was required. But the fate of unobserved quantum states remains unresolved.

Modern interpretations include decoherence, the idea that quantum possibilities become eroded from interactions with the classical world, for example thermal energy or heat. Efforts to build quantum computers must contend with decoherence as the enemy to be avoided long enough for quantum computations to be completed. Some technologies use extremely cold temperatures to prevent decoherence. But the issue of isolated, large-scale quantum states persists. Most importantly, increasing evidence in recent years shows that living systems at warm temperatures avoid decoherence and utilize quantum coherence and entanglement.

Other interpretations include the multiple-worlds view, in which each and every quantum possibility branches off to its own new universe, resulting in an infinity of overlapping worlds. Others propose an objective threshold—a critical level at which quantum systems spontaneously reduce, or collapse, to definite classical states, preserving one universe.

One such proposal for an objective threshold for reduction ("objective reduction," a.k.a. "OR") was put forth by British physicist Sir Roger Penrose in 1989, and includes several features relevant to the conscious mind. In a bold and profound assertion, Penrose tied consciousness to the deepest level of the universe.

He began by approaching the nature of qwffs surrounding how a particle can conceivably exist in multiple locations or states simultaneously. Einstein's general relativity equated matter with curvature in spacetime. Penrose reasoned that a particle in two places at the same moment—a particle separated from itself—implied that *spacetime* separated, e.g., curved in opposite directions simultaneously. Qwffs may be viewed as separations, blisters, or wavelike ripples in the fabric of the universe.

In the multiple-worlds view, each separated spacetime possibility of a qwff forms its own universe. But Penrose concluded such qwffs/separations are unstable and due to an inherent feature in spacetime would reduce, or self-collapse upon reaching threshold for objective reduction (OR). Waves would self-collapse, not unlike ocean waves crashing on a beach. Each instance of this type of self-collapse, according to Penrose, was associated with a moment of conscious awareness.

But what does it mean for the universe to separate, to bifurcate? What is the universe made of? Matter is composed of atoms, but atoms are mostly empty space, as is the space between atoms. The universe is mostly empty. But what is curving? What separates from itself? Einstein called it the spacetime continuum. It's everywhere, but . . . what is it? *Where* is it?

It's at the Planck scale, the ground floor of the universe. Imagine you were able to shrink smaller and smaller while everything else remained the same size. Let's say that every minute you became ten times smaller than you were the minute before. In about three minutes you'd be the size of an ant, in another three the size of a living cell, another three the size of an atom, and three more the size of subatomic particles like electrons and quarks.

As you get increasingly smaller, matter is gone; just featureless emptiness remains. But keep shrinking and shrinking for another twenty minutes or so, and patterns begin to appear, irregularities in the world around you. (Physicist John Wheeler compared this to viewing an apparently smooth ocean surface from outer space, but seeing roiling waves and whitecaps when closer.) At the fine grain, the infinitesimal Planck scale, the universe is no longer smooth but quantized, composed of geometric patterns termed fundamental spacetime geometry.

Science cannot yet observe Planck-scale spacetime geometry, and may never. But fields like string theory and quantum gravity portray the fine grain of the universe as a 4-dimensional evolving web of various Planck-scale volume pixels of spin. The patterns seem to repeat periodically over distance and scale, suggesting the universe is, at this level, holographic, possibly accounting for nonlocal entanglement.

Are the patterns random? Or could actual information be embedded there? In another bold assertion, Penrose proposed the latter: that embedded in patterns of Planck-scale spacetime geometry were Platonic values, including mathematical truth, aesthetic, ethical, and proto-conscious values. Although Penrose avoids such comparisons, his views match Buddhist, Vedantic and Kabbalah beliefs about an infinite "emptiness" of consciousness and wisdom.

Qwffs may be seen as spacetime separations, ripples in the universe at the level of Planck scale geometry. With Penrose OR, ripples become self-organizing, and connected to Platonic values and the raw components of conscious experience. Like waves on a beach, qwffs self-collapse in sequences of conscious moments.

When Penrose introduced OR as a mechanism for consciousness, he didn't have a biological structure for quantum computing in the brain, particularly one able to avoid decoherence. When I read his ideas in the early 1990s, I had for twenty years studied computational capabilities of protein lattices called microtubules inside neurons. They were ideal candidates for biological quantum computing. Penrose and I developed a model of OR-mediated quantum computation in microtubules inside brain neurons. Synaptic inputs were proposed to "orchestrate" quantum computations, hence the neurobiological model became known as "orchestrated objective reduction" ("Orch OR"). Critics claimed decoherence in the warm brain precluded Orch OR, but evidence in recent years clearly shows quantum coherence in warm biological systems.

Penrose used one simple equation to describe a threshold for conscious moments, one in which the qwff spacetime ripples and the time to reach threshold are inversely related. This defines a spectrum of conscious moments, consistent with Buddhist and Vedantic beliefs. The human brain has typically forty conscious moments per second. Heightened states, e.g., in meditating Tibetan monks, reach up to 100 Hz, with high amplitude and coherence. The upper limits of conscious experience remain unexplored.

Consciousness may be seen as a process, sequences of self-organizing waves/qwffs rippling on the edge between quantum and classical worlds, tied to fundamental spacetime geometry submerged at the Planck scale. With entanglement, nonlocal interactions among living beings and physical events can in principle occur (the "Law of Attraction"). Shaping one's reality, psychic phenomena/ESP, influence by Platonic values (or "following the way of the Tao," or "divine guidance") and even life-after-death become plausible. This is not to say they do occur, but simply that they are not at all impossible.

How could Platonic values and conscious components have become embedded in Planck-scale geometry? During the Big Bang, the universe expanded very rapidly for an extremely brief period of time (so-called "inflation") and then settled down to slow, gradual expansion continuing to this day. Physicist Paola Zizzi suggested that during rapid inflation the universe was in a quantum qwff state of multiple possible universes. She calculated the end of inflation coincided with threshold for Penrose OR for the universe, suggesting the Big Bang included a cosmic conscious moment (the "Big Wow"). So we all just may be ripples in a cosmic conscious sea.

—Stuart Hameroff
September, 2009